Utilization of Agricultural Waste for the Production of High Valued Products

AF773551

NIPA® GENX ELECTRONIC RESOURCES & SOLUTIONS P. LTD.
New Delhi-110 034

About the Author

Dr. Renu Agrawal is Ex. Chief Scientist, CSIR-CFTRI and Rural Development Programme Coordinator, Mysore. She is a science columnist, reviewer, science reformer, counsillor, author of books and works for social upliftment. She has represented India as a team leader taking scientific delegation for Asia meet at Indonesia, Bali nominated by DST, Government of India. She had her PhD from University of Rajasthan, Jaipur. Her areas of specialization include biotransformation and probiotics. She has received many externally funded project grants as principal investigator from DBT (Department of Biotechnology) and DST (Department of Science and Technology) Govt. of India and Bilateral International programmes with Bulgaria and Argentina. Her path breaking and innovative research is evidenced by her publications. She has developed many innovative functional foods and technologies oriented towards improving health by natural means. The survey shows that these technologies are the most sought of and downloaded by more than 1900 people and are used by the common man. She has guided students for M.Sc, M. phil and PhD in Biotechnology and Microbiology. She has published more than 60 research papers in peer reviewed national and international journals. She has presented more than 160 papers in various national and international conferences and delivered more than 300 talks. She has 20 patents to her credit. She has received **Karamveer Chakra from Rex-UN global, 2016**, **"Äbdul Kalam Life Time achievement National Award"**, 2017, by International Institute of Social and economic Reforms in presence of Union Minister Mr. Sadananda Gowda for her untiring work for the development of the nation and outstanding excellence in the field of Microbiology and Biotechnology. She has been bestowed with **Kissan Ratna Puraskar by Junior Chamber International** in 2017. She has received the **Nari Shakti** award, 2018, from Social reforms research organization with Ministry of women and child development, Govt. of India.

She has been honored as **Fellows** of "Fellow of Association of Microbiologists of India", Fellow member of International Society of Biotechnology', Fellow of National Academy of Biological Sciences and "Fellow Society of Applied Biotechnology". She has bagged the **"Women Achievers award'** by the "International Guild of Women Achievers" and India International Friendship society. She has been selected as the **"Best woman scientist"** by National Academy of biological sciences, felicitated by Rotary Inner wheel for scientific excellence with **"Suguna Award"** and felicitated by Institution of Engineers, Mysore unit. She has got **'Reflections award"** by Rotary Inner Wheel for doing most outstanding projects for the welfare of the society. She is the editor of Research Journal of Biotechnology, ADC, Biostandups, Journal of natural products and Associate Editor of International Journal of Agriculture Sciences. She is a reviewer of many national and international Journals. She has authored a book on Probiotics and many book chapters including from Marcell Dekkar, USA and Taylor and Francis, UK. Her work has won many best paper awards at the national and international conferences. She has taught as faculty of M.Sc Food Technology at CFTRI and to ACSIR students. She has served as the president of 'Association of Microbiologists of India", Mysore chapter and has been a member of many international Technical Advisory Boards. She has been on the National committees of NCERT of national Science Talent Search, scientist selection boards of UPSC, DST and DBT and many international technical committees at Dubai, Bulgaria, Argentina and Indonesia. She is an Executive committee member of "National Academy of Biological Sciences".

In Rotary Inner wheel she has been awarded the **Best secretary** for two times and also **Best ISO** twice in the district 318. She has been serving the society by her work. She is a philanthropist and has been running many fellowships and national level competitions for poor meritorious students, constructed pre fixed auto stand in Mysore at Hardinge circle for public use. She has donated an Electrolyte for rapid detection of HIV to AshaKirana hospital, Mysore.

She has been nominated to Co-Chair the Expert Committees on Scheme for Young Scientists and Technologists (SYST), Department of Science and Technology, Govt. of India. It plays an important role in nurturing young scientists with their ideas to undertake projects of societal relevance and selection of the same. She has been nominated as a member of National Experts Advisory Committees (NEAC) from Government of India, Ministry of science and Technology for evaluating and monitoring various programmes involving children, Children science congress, innovative research and outreach, ASAR-CODER, projects to be taken in Desert areas, "Pandit Deen Dayal Upadhyay Gram Pariyojna" and in 'KIRAN" which involves Indo –US projects for women. She writes regular column under science scope in City Today" newspaper. She has been honored as **"SHERO" by Biostandups**, an online science magazine. She has been the opening year (January, 2017) Face of the Month for the magazine. She has been honored to be on their editorial board and writes articles for them. She reviews many scientific papers every month" for national and International journals. She is also involved in the upliftment of rural education of Karnataka through "Pratham" and HRDC, University of Mysore through teachers training workshops and in translating NCERT course book material to be made into short videos to make science easy and fun for the children.

Her commitment to work can be seen as she was taken as Project Advisor by CFTRI even after her retirement.

Utilization of Agricultural Waste for the Production of High Valued Products

Renu Agrawal

Ex. Chief Scientist, Food Microbiology
CSIR-800, Rural Programme Coordinator
Director, S&T Unit
Central Food Technological Research Institute
Cheluvamba Mansion, Opp. Railway Museum
Mysuru, Karnataka - 570 020, India

NIPA® GENX ELECTRONIC RESOURCES & SOLUTIONS P. LTD.
New Delhi-110 034

NIPA® GENX ELECTRONIC
RESOURCES & SOLUTIONS P. LTD.

101,103, Vikas Surya Plaza, CU Block
L.S.C. Market, Pitam Pura, New Delhi-110 034
Ph : +91 11 27341616, 27341717, 27341718
E-mail: newindiapublishingagency@gmail.com
web: www.nipabooks.com

For customer assistance, please contact
Phone: + 91-11-27 34 17 17
Fax: + 91-11- 27 34 16 16
E-Mail: feedbacks@nipabooks.com

ISBN: 978-93-58870-48-0

Composed and Designed by NIPA®.

Dedicated to
Farming Community

Kerala Agricultural University

Main Campus, Vellanikara, KAU, PO, Thrissur - 680 656, India
Phone: 0487-2373017; www.kau.in

Foreword

The book "Utilization of agricultural wastes for the production of high valued products" authored by Dr. Renu Agrawal Formerly at C.F.T.R.I Mysore aims to create awareness on utilization of wastes produced in agriculture, horticulture, forestry and allied areas of livelihood. Agricultural wastes are the residues from the raw agricultural products like fruits, vegetables, meat, poultry, dairy products and various crops. The processing of agricultural products benefits man from time immemorial. In recent times a lot of innovations have taken place in the field of agriculture. It is important that for the empowerment of farmers, the farmers must align themselves with the agricultural scientists.

Farmers face a lot of problems mainly due to shortage of storage space. The farmers should be empowered to convert their produces into high valued products. At the same time they should know how to utilize the wastes produced during agriculture. This will help not only the farmers but the community at large in keeping the environment green and improving human health.

The present book has been divided into seven chapters. **Chapter 1** deals with the introduction on what is the necessity for the utilization of agricultural wastes. **Chapter 2** deals with different types of agricultural wastes and tips to transform them into effective resources which could be utilized towards sustainable development. **Chapter 3** deals with the different types of agro-industrial wastes, the sources and the problems associated with agricultural wastes. It also deals treatment of these wastes and how the wastes could be disposed using clean technology(s). **Chapter 4** deals with production of useful products from agro wastes. It also deals with the reuse, recovery and recycling of the wastes. The subsection covers the production of important products like biofuel, enzymes, fiavonoids, anthocyanins, flavones, alkaloids, terpenoids, steroids, phenols, essential oils, auto-inducers, phenazines, polypetides, peptides, bioeneregy, nano materials, compost and microbial bioconversions from agricultural wastes. **Chapter 5** deals with novel approaches for natural products biogeneration by strain improvement, genetic improvement, mutations, protoplast fusion and molecular genetic tools. It is very important that after the products are produced they should be commercialized. **Chapter 6** deals with the commercial utilization of the products produced and deals with the future implications. Farmers are expected to produce more food per unit of land, water and agro chemicals. There are many challenges to be faced like climate change, volatility and change in nutritional requirement etc. It is observed that in recent times the public looks forward to tailor made specific functional foods. Therefore the task of the farmer has become very challenging. The chapter also deals with the benefits to the food industries and the consumers in general and what will be the future implications and the vision in this direction.

This book provides a very comprehensive review starting from introduction on agriculture wastes to the future implications including novel foods and nutrition security. The book provides the most recent updates in the area of food processing with a holistic approach.

Dr. Renu Agrawal is the epitome of food science in India and a Fellow of science academies including National Academy of Biological Sciences. A well known and dedicated science communicator, she specialized in value chain in food industry. Author of several research and extention articles she has authored this book for the benefit of students, teachers, scientists and policy planners.

The New India Publishing Agency (NIPA) New Delhi is a well reputed Indian Publisher and I congratulate the author and the publisher for a very valuable and educative book.

K.V. Peter

Preface

Agricultural wastes are obtained after the production and processing of agricultural products. It is possible that these can benefit the human mankind. The economic values should be more than the cost of collection, transportation and processing for beneficial use and should be economically feasible. In Agriculture many wastes are obtained. The use of chemicals in agriculture also affects the environment which causes a lot of pollution. The wastes generated from crop have a good potential to convert into energy or can be converted into useful products. The waste produce from animal waste and crop residues has many physicochemical properties. The present book deals with the research work carried out in the past related to conversion and utilization of agricultural waste into useful products and to increase their economic values. The book mainly aims in analysing the various application and research carried in the fields of potential utilization of agricultural wastes. The recycling and utilization of agricultural wastes is an important step forward towards environmental protection, energy structure and agricultural development. The recycling and utilization pathway of agricultural wastes have also been discussed. The book also deals with the laws and regulations and strengthening of rural market. It will provide more comprehensive fundamental information for the recycling and utilization of agricultural wastes during the modernization and urbanization around the globe.

Author

Contents

1

Introduction

In today's times, a lot of innovation has come into agriculture. This very hard occupation requires an amalgation of farmers with the scientists. From the rural areas after education the technocrats are moving to the cities and this is creating a great loss to the agriculturists. As the population is growing so is the demand for food. Due to shortage in the storage systems the amount of wastage has increased a lot. It is an important aspect to be looked into. A lot of waste material is seen in the vegetable markets, city markets and near the slums. This causes a lot of pollution and spoils the environment. It is important to know why is it increasing and why the facilities for wastes do not exist. It is also seen that the nutrients leach out from the damp fills which make the rivers polluted. This enhances the growth of water weeds. Therefore, if public could be made aware as to how they could utilize the wastes which could then be converted into valued resources improving the environmental pollution and human health.

2

What is Agricultural Waste

After the agricultural activities a number of by-products are obtained and these are known as agricultural waste. These include residues of different crops like stalks, straws, leaves, roots, husks and shells. The animal waste is in the form of manure. All these agricultural wastes are free of cost however; they can be utilized as an important resource (Lionel and Arnold, 1931). Agrawal (1998) has mentioned India as the final dumpyard. The wastes obtained from various resources can be converted into heat, steam, charcoal, methanol, ethanol, biofuel, compost, energy, biogas, construction and high valued chemical compounds. Many times these agri-industrial wastes are not completely utilized and either they rot in open areas or are burned in the fields. Singh and Nain (2014) studied the role of microorganisms in the conversion of agricultural wastes to compost. These wastes are found to be rich in minerals like nitrogen, potassium and phosphorus. These help in improving the soil fertility. It has been obsereved that as the soil fertility enhances, the yields of the crops also increase. This in turn is beneficial for food security and environmental health and also in managing the utilization of organic wastes in agriculture. Alburquerque *et al.*, (2006a) studied the detoxification and maturity in compost made from "alperujo", the solid by-product of extracting olive oil by the two-phase centrifugation system. Frontera-Suau *et al.*, (2002) studied the aerobic biodegradation of hopanes and other biomarkers by crude oil-degrading enrichment cultures. Somshekar *et al.*, (1999) prepared a novel enrichment technique for the isolation of highly potent catalase producing yeast from soil. Bioconversion of agro-industrial wastes for the production of fibrinolytic enzyme from *Bacillus halodurans* IND18 was studied by purifying and characterizing biochemically (Vijayaraghavan *et al.*, 2016). Wu *et al.*, (2005) studied the influence of fertilizers and organic amendments on organic carbon fractions in Heilu soil on the loess plateau of China. The wastes could be useful for the production of value added products as improved feed, bio-processing reagents with selective catalysts, safe green chemicals, bio-fuels, biogas, bio-plastics and biopolymers.

This can serve the agro-processing sector in and around the region by making it more resource efficient and sustainable. This is very crucial for the farmers and it benefits them. Increase in the demand for local crops and bio-resources helps in rural livelihoods. It also enhances the agribusiness opportunities for farmers and small scale enterprises in the region. Import of expensive fossil fuel can be reduced by promoting the conversion of waste into renewable energy as biogas and bio-fuel. This will also reduce the need to mitigate climate change. Singh *et al.*, (2012) have studied the role of microorganisms in sustainable agriculture and biotechnology. Singhal and Pandey (2001) studied the solid waste management in India. The status and future directions. Smolinska (2000) studied the survival of sclerotium in soil with cruciferous residues.

This book has been written with an objective to review the studies done to make the agricultural wastes into very effective resources that can be utilized and not discarded. Governments can set up various national institutes that will utilize these potential of converting agricultural wastes and convert them into high valued products. Le Blanc (2015) has written a very good article on the sustainable development goals as a network of targets towards sustainable development.

3

Different Types of Agricultural Industrial Wastes

3a. Sources and problems of agricultural wastes

3b. Waste Reduction

3c. Waste Treatment

3d. Waste Disposal

3e. Clean Technology

Different Kinds of Agricultural Wastes

Table 1: Common Hazardous Wastes: Community Source

Common Hazardous Wastes	Community Source
Explosives	Construction companies, dry cleaners, ammunition production facilities, etc.
Radioactive substances	Biomedical research facilities, colleges anduniversity laboratories, offices, hospitals, nuclear power plants, etc.
Biomedical research facilities, colleges and	Biomedical research facilities, drug companies, hospitals, medical clinics, etc.
Flammable wastes	Dry cleaners, petroleum reclamation plants, petroleum refining and processing facilities, service stations, tanker truck cleaning stations, etc.
Nuclear power plants, etc.	
Toxic chemicals	Agricultural chemical companies, battery shops, car washes, chemical shops, college and university laboratories, construction companies, electric utilities, hospitals and clinics, industrial cooling towers, newspaper and photographic solutions, nuclear power plants, pest control agencies, photographic processing facilities, plating shops, service stations, etc.

Source: Tchobanoglous, *et al*. (1993).

Table 2: Wastes of Plastic resin generated and Recycled

Resin	Code	Amount generated (thousand tons)	Amount recycled (thousand tons)
Polyethylene terephthalate	1	2860	540
High-density polyethylene	2	5890	520
Polyvinyl chloride	3	1640	
Low-density polyethylene	4	6450	190
Polypropylene	5	4000	10
Polystyrene	6	2590	
Other resins	7	5480	390

Source: American Plastics Council (2006b) and EPA (2006a).
"Includes linear low-density polyethylene.

Table 3: Country Based wastes and energy Recovery.

	Total (million tonnes)	Landfill	Incineration	Energy	Backfilling recovery	Recycling
EU-28	2319.5	47.4	1.5	4.7	10.2	36.2
Belgium	42.8	8.2	4.3	13.6	0.0	73.9
Bulgaria	175.7	97.9	0.0	0.1	0.0	2.0
Czech Republic	19.9	17.3	0.4	5.1	29.1	48.1
Danmark	17.7	21.7	0.0	20.7	0.0	57.6
Germany	370.7	19.2	2.3	10.5	25.3	42.7
Estonia	20 .7	65.6	0.0	2.5	11.9	20.0
Ireland	10.0	42.6	0.1	7.2	37.4	12.7
Greece	67.1	88.4	0.0	0.2	8.1	3.2
Spain	103.4	47.9	0.0	3.4	12.6	36.1
France	299.7	29.3	2.0	4.5	10.7	53.6
Croatia	3.5	51.1	0.0	1.4	2.0	45.5
Italy	129.2	16.0	5.2	1.6	0.2	76.9
Cyprus	1.8	58.9	0.0	1.7	25.9	13.5
Latvia	1.9	34.8	0.0	8.7	0.9	55.5
Lithuania	4.5	67.6	0.1	4.1	2.5	25.8
Luxembourg	6.5	38.3	0.0	2.5	16.0	43.3
Hungary	13.7	39.4	0.7	8.9	3.7	47.3
Malta	1.6	38.9	0.4	0.0	37.5	33.3
Netherlands	130.6	45.4	1.0	7.9	0.0	45.7
Austria	53.9	38.6	0.2	6.5	20.1	34.7
Poland	182.4	24.9	0.4	2.7	21.5	50.5
Portugal	9.9	31.8	10.0	3.1	0.0	55.0
Romania	172.2	94.4	**0.0**	1.3	0.6	3.7
Slovenia	5.4	9.2	0.6	4.9	33.5	51.8
Slovakia	7.1	53.8	0.8	4.4	0.0	40.9
Finland	93.3	30.9	0.5	4.8	**0.0**	13.8
Sweden	183,3	84.4	0.1	4.7	1.6	9.3

United Kingdom	209.0	41.5	3.6	0.9	10.4	43.6
Iceland (')	0.5	30.7	0.0	2.7	0.6	66.0
Norway	11.7	17.9	0.5	35.8	5.3	40.5
Montenegro	1.0	98.8	0.0	0.1	0.0	1.0
Former Yugoslav Republic of Macedonia	1.5	98.7	1.3	0.0	0.0	0.0
Albania	1.2	74.8	3.1	0.5	0.0	21.6
Serbia	49.4	97.3	0.0	0.1	0.0	2.6
Turkey	79.3	70.2	0.0	0.7		29.0

(') 2012. *Source*: Eurostat (online data code: erw_wastri)

Source: Tchobanoglous, *et al.*, (1993)

Agricultural wastes in the form of crop residues or animal manures could be converted into organic fertilizers. This will not only reduce waste, clean the environment but will also kill the pathogenic microorganisms and will inhibit the germination of weeds in the fields. Kobayashi *et al.*, (2009) characterised solid residues which were obtained from hot-compressed-water treatment from woody biomass. Different kinds of agricultural wastes are given in Table-1.

In an awareness programme among farmers (120 farmers in India) for the utilization of agricultural waste like that of biogas plant waste, mushroom waste, wheat waste, mustard and horticultural waste was found to be more than 70 percent. Thus, overall awareness about utilization of agricultural waste was very high but utilization of agricultural waste by the farmers was very less. Benefits were seen in the form of high milk yield, more income by selling waste to the brick yards, reduced expenditure on chemical fertilizers, reduced waste available for disposal or clean and safe environment. Government's can establish waste collection centers in villages which will make it easy for the villagers for proper disposal (Table 2, 3).

When processing units are set up in rural areas it will increase the employment opportunities for the rural folk to make value added products. Establishment of handicraft industries (for making bags, mats, hats, carpets and handmade paper) are a good preposition. A small scale processing unit for the utilization of paddy straw can be very useful. This can be used as a source of energy. Mushroom processing unit can be easily run by village farmers utilizing biomass energy obtained from wastes like paddy straw, cotton sticks, mustard sticks or husks. Gaur (1999) has worked out on a microbial technology for composting the agricultural residue by an improved process. Jain and Vyas (2002) studied the yield response of *Pleurotus florida* on heat straw in combination with other substrate. Zhang *et al.*, (2010) overviewed the recent advances in thermochemical conversion of biomass. Agrawal and Basappa (1994) studied the fermentation of paddy malt mash to ethanol by mixed cultures of *Saccharomyces cerevisiae* and *Zymomonas mobilis* ZM4 with Penicillin G. Agrawal and Basappa (1996) studied the role of antimicrobial agents in simultaneous saccharification and fermentation of paddy malt mash to ethanol by mixed cultures of *Saccharomyces cerevisiae* PHO3 and *Zymomonas mobilis* ZM4.

Wheat and Paddy waste: A byproduct of wheat crop is straw which is utilized for making many products. However, many farmers use the mas animal feed. Many farmers are aware of the utility in making particle board but they do not sell wheat straw to particle board industries or utilize it in products like briquettes, dry flowers, hats, mats, carpets and other handicrafts. Garay *et al.*, (2009) were the first to find out that it was possible to make particle board with crop residues along with wood obtained from Pinus. The best waste found was of lignocellulosic material. It was found (Cruz *et al.*, 2001) that lignocellulosic materials are good sources of antioxidant compounds. This is made up of cellulose, hemicellulose and lignin in different concentrations depending on the nature and source of the biomass. In lignocellulosic waste the precursor lignin is converted into phenolic compounds. Lignin is made up of heterogeneous polymer which is phenolic in nature. This contains precursors like transconiferyl, trans-sinapyl and trans-p-coumaryl alcohols. In Gymnosperms mainly guaiacyl groups are present in lignin. However, in woody angiosperms lignins contain guaiacyl–syringyl groups and lignins from grasses contain guaiacyl–syringyl-phydroxyphenyl groups. Ando *et al.* (2000) found that during the hydrolytic processing lignin and lignin–hemicellulose linkages are partially depolymerized. In arabinoxylans or pectins the ferulic and p-coumaric acids are found to be linked by ester bonds. There are condensed tannins like proanthocyanidins and hydrolyzable ellagitannins which are present in the phenolic fraction of hard wood (Cadahýa *et al.*, 2001).Vicuña (1988) studied the bacterial degradation of lignin. Doughari *et al.*, (2008) studied the antioxidant activity of lignin phenolic compounds which were extracted from kraft and sulphite black liquors. Ball *et al.*, (1989) studied the degradation of lignin-related compounds by actinomycetes. According to Conde *et al.*, (1995) acids (gallic, vanillic and ellagic) and aldehydes (syringaldehyde and sinapaldehyde) are also present. From hydrolysates the phenolic compounds can be recovered after extraction with solvents like ethyl acetate or diethyl ether. Ethyl acetate helps in removing the water-soluble phenols and also hemicellulose degradation products. The aqueous phase contains the lignin–carbohydrate complexes (Bouchard *et al.*, 1991). Akanbi and Togun (2002) have studied the influence of maize–stover compost and nitrogen fertilizer on growth, yield and nutrient uptake of amaranth.

Angerosa *et al.*, (1995) evaluated the phenolic compounds in virgin olive oil by GC-MS. Visioli and Galli (2002) studied the biological properties of olive oil phytochemicals. Capanema *et al.*, (2005) characterized quantitatively the hardwood milled wood lignin by nuclear magnetic resonance spectroscopy. Chen *et al.*, (1977) surveyed and have provided information on humic substances. Termorshuizen *et al.*, (2007) studied that 18 composts were suppressive against 7 patho systems and had variability in pathogen response. Chenoweth *et al.*, (1975) studied the metabolism and physiological effects of pectins. Ehara and Saka (2005) studied the behavior of cellulose decomposition in supercritical water, subcritical water and their combined treatments. Ehara and Saka (2002) studied a comparison on the chemical conversion of cellulose between the batch-type and flow-type systems in supercritical water. Soundar and Chandra (1987) studied the production of cellulase and detection of avicel-adsorbing carboxy methyl cellulase from a mesophilic fungus. Gaind and Nain (2010) explored the composted cereal waste and poultry manure for the restoration of soil. Gaind *et al.*, (2009) evaluated the quality of co-composted wheat straw, poultry droppings and oil seed cakes. Gajalakshmi and Abbasi (2008) have written an excellent review on the solid waste management by composting. Planning Commission, Government of India (2014) has come out with a report of the task force on waste to energy (I) in the context of integrated municipal solid waste management. Sharholy *et al.*, (2005) have analysed municipal solid waste systems in Delhi. Godden *et al.*, (1992) elucidated the lignin degradation pathway in actinomycetes. Hamelinck *et al.*, (2005) produced ethanol from lignocellulosic biomass using techno-economic performance which could be helpful for short, middle and long-term. Venkateswarlu and Rao (1983) studied the response and reduction of pearl millet to inoculate with different strains of *Azospirillum brasilense*. Scott *et al.*, (1949) studied the partial purification and properties of antibiotic substances from the banana (*Musa sapientum*). Sluiter *et al.*, (2005) determined the structural carbohydrates and lignin for biomass formation. To measure aquatic humus content work was done by analysing the samples by spectroscopic analysis (Hautala *et al.*, 2000). Janshekar and Fiechter (1983) studied the aspects of lignin biosynthesis, applications and its biodegradation. Kanotra and Mathur (1994) conducted biodegradation of paddy straw with the help of cellulolytic fungi and applied it on the wheat crop. Khalid *et al.*, (2008) conducted a comparative study of polypropylene composites which was reinforced with oil palm empty fruit bunch fiber and oil palm derived cellulose. Rodriguez *et al.*, (2008) used high-boiling point organic solvents for pulping oil palm empty fruit bunches. Rozman *et al.*, (2003) studied the flexural and impact properties of oil palm empty fruit bunch (EFB)–polypropylene composites–the effect of maleic anhydride chemical modification of EFB. Rozman *et al.*, (2001a,b) studied the tensile properties of oil palm empty fruit bunch–polyurethane composites. Tartoura and Youssef (2011) stimulated ROS-scavenging systems in squash (*Cucurbita pepo* L. var. Hybrid fadwa) plants with the supplementation of compost under normal and low temperature conditions. Vasudevan and Rajaram (2001) studied the bioremediation of oil sludge-contaminated soil.

Saritha *et al.*, (2012) have shown how the paddy straw can be pretreated with *Trametes hirsute* for improved enzymatic saccharification. Shymal *et al.*, (2011) studied the yield, fruit body diameter and cropping duration of oyster mushroom (*Pleurotus sajor* caju) grown on different grasses and paddy straw as a substrate. Sidhu *et al.*, (1998) studied the sustainability implications of burning rice and wheat straw in Punjab. Kirk (1983) studied the methods of degradation and conversion of lignocelluloses. Kumar *et al.*, (2008) evaluated thermophilic fungal consortium for composting paddy straw. Pandey *et al.*, (2009b) investigated the effect of bio augmentation and nitrogen supplementation on composting of paddy straw. Pant *et al.*, (2006) cultivated oyster mushrooms on wheat straw and bagasse as substrate with distillery effluent. Sidhu and Beri (1989) have written an excellent article on the effect of crop residue management on the yields of different crops and on soil properties. Suman and Sharma (2005) studied the cultivation of mushrooms and its uses. Tanaka *et al.*, (2007) utilized lignocellulosic biomass which was discharged from palm oil industries. Yang *et al.*, (2006) studied the pyrolysis of palm oil wastes to enhance the production of hydrogen rich gases. Rama *et al.* (2015) made an assessment study for solid waste management system in Chandigarh city of India.

3. a. Sources and Problems of Agricultural Wastes

The Sugarcane waste:

From sugarcane crop the wastes obtained are sugarcane trash and bagasse. A survey was done with sugar cane farmers where 50% were using bagasse as fuel in making jaggery, 46.51% converted it into compost or vermicompost. It was only 11.62% who sold it to paper/cardboard industry. All analysis was done by proper methods given in A.O.A.C. (1975). Laser *et al.*, (2002) studied a comparison of liquid hot water and steam pretreatments using sugar cane bagasse for the bioconversion to ethanol. Sasaki *et al.* (2003) fractionated the sugarcane bagasse by hydrothermal treatment. Yan *et al.* (2009) worked on the thermal pretreatment of lignocellulosic biomass.

Another advantage is in growing plants from green fodder. Production of biogas is also done by sugarcane bagasse. Similar work was done by Dellepiane *et al.*, (2003) who conducted the study due to the difficulty in finding energy sources and in order to reduce pollution. In particular, the biogas comes from bagasse and barbojo, which is a sugarcane residue. In the developing countries most of the times it has been observed that it is wasted or sometimes utilized with inefficient technology. Clemente *et al.*, (2010) studied the trace element mobility in soil contaminated for two years after field-amendment with green waste compost mulch. Clemente *et al.*, (2005) studied the way heavy metals were taken up by *Brassica juncea* grown in a contaminated soil in Aznalcóllar (Spain). Karaca *et al.*, (2006) studied the effects of a humic deposit (gyttja) on soil chemicals and investigated the microbiological properties and heavy metal availability. Dellepiane *et al.*, (2003) studied the feasibility of obtaining clean energy from sugarcane waste. They have shown an innovative application of bagasse and barbojo.

Cotton waste

Cotton sticks are usually left after the picking of cotton which can be used as a fuel. The farmers can easily sell these waste cotton sticks to power plants, plywood industries or to particle board industries which may add to their income. It can also be utilized by chipping and converting them into compost or can also be used in biogas production after treating it an aerobically. It was Isci *et al.*, (2007) who found that cotton wastes are a good source of biogas. From 1 g of either of cotton stalks, cotton seed hull or cotton oil cake around 65, 86 and 78 ml of CH_4 can be produced in 23 days.

Mustard waste

Two major by products of mustard crop are mustard sticks and husk. These are sold to the brick industries by the farmers who cultivate it. Most of the farmers store them for future use. Some of them also use them for burning in chulha. However, it is rarely used for composting. Managing mustard sticks by chipping and composting or feeding the animals after the treatment with ammonia has been found to be profitable.

Many important materials can be made from mustard sticks which include briquettes. Mustard stalk, mixed waste of tree leaves and grasses (3:1) and wood waste along with three organic binding materials (molasses, press mud and distillers dry grain) with varying concentrations of 5, 10, 15 and 20% can be used for the preparation of briquettes. Press mud is found to be a good binding agent. This is followed by distiller's dry grain and molasses. Andrade *et al.*, (2001) found that the physical characteristics like moisture content, bulk density, compression ratio and compressive strength were required for better utilization and safe handling during transportation at a pressure of 123.42 MPa. Bio-oil has also been derived from empty fruit bunches (Abdullah *et al.*, 2008).

Alam *et al.* (2007) have successfully derived activated carbons from empty-fruit bunches of oil palm. This will be helpful in solving environmental problems.

Alburquerque *et al.*, (2006a) have measured the levels of detoxification and maturity in compost made from "alperujo", the solid by-product of extracting olive oil by the two-phase centrifugation system. Alburquerque *et al.*, (2006b) studied the effects of bulking agent on the composting of "alperujo", the solid by-product of the two-phase centrifugation method for olive oil extraction.

Horticultural waste

Among the horticultural crops, wastes usually include damaged or spoiled fruits and vegetables, dead plants, branches, leaves and unsold fruits and vegetables. These can be converted into compost/vermicompost or fed to animals by more than 70.31 percent of the farmers. Value added products can be made from surplus fruits and vegetables and then sold into the market. This helps the farmers in avoiding wastage and helps them to earn more. Preservation of unsold fruits and vegetables could be done by drying them and then selling. Useful chemicals can be extracted from waste fruits and vegetables which include products like citric acid, lactic acid, acetic acid and others. Zhang *et al.*, (2011) studied the extraction of pectin from potato residues. Dhillon and Kaur (2016) have studied the agro-industrial wastes to be used for enzyme production. Frutos *et al.*, (2008) worked out the strategies for good fertilizers for field vegetable production.

Floricultural waste

After flower picking, the whole plant becomes a waste. This can be used for composting, vermicomposting and green manuring. The dead plants and waste flowers can be used in composting, vermicomposting or in green manuring. The left out flowers can be dried and powdered. The cut flowers can be used for making dry flowers which is an upcoming industry. The dry-flowers can be painted with different colors and dyed. Many floral products like cards, pictures, wall hangings and potpouris can also be prepared from them. Dilta *et al.*, (2011) have reported similar uses of dry flowers.

Mushroom waste

Mushrooms are produced on natural agriculture materials like wood, animal husbandry or wastes from manufacturing industries. After mushroom crops are harvested, millions of tons of "spent" (used) mushroom substrate become available for other uses. This can be used for various products as manure, for gardening, for making nurseries or growing vegetables by the farmers. It can also be used as animal feed or for biogas generation. Sánchez *et al.*, (2002) studied the handling as a component in potting soil mixes, in agriculture or in landscapes in order to enrich the soil. It can be used as a casing material, in vermiculture as a growing medium, for remediation of contaminated water in wet lands, as fuel and also as medium for vermiculture. It can also be used to stabilize disturbed soils which can be used as an animal bedding or animal feed and to inhibit plant diseases. Rinker and Alm (2002) evaluated spawn and its role in green mold disease. Sánchez *et al.*, (2002) have written a very informative book on mushroom biology and mushroom products. Sing *et al.*, (2003) have written a good article on the nutritional composition, processing and preservation of the edible mushroom found in Manipur using certain agro-waste and forest by product for sustainable development. Singh *et al.*, (2012) have studied the role of microorganisms in sustainable agriculture and biotechnology. Smolinska (2000) studied the survival of *Sclerotium cepivorum* sclerotia and *Fusarium oxysporum* chlamydospores in soil amended with cruciferous residues. Trillas *et al.*, (2006) studied the composts from agricultural waste and the feasibility of *Trichoderma asperellum* strain T-34 to suppress *Rhizoctonia solani* in cucumber seedlings.

Wastes from animals like faeces, urine and fermentation gases are of great concern around the world. Moreover, as the excretion is in different forms like solid, liquid or gases it causes a serious problem. Odour pollution is another major problem among the population mostly seen as nitrogen pollution. When the manure decomposition takes place it releases methane (CH_4), ammonia (NH_3) and nitrogen oxides which results in high levels of greenhouse gases. When ammonia volatilizes it causes acid precipitation. Similarly, nitrification and denitrification causes ozone depletion. Garima and Singh (2014) have reviewed studies on bioremediation on solid waste management and how they could be applied. Strom (1985) has identified some thermophilic bacteria for composting the solid-waste. Madan *et al.*, (1987) cultivated *Pleurotus* on different waste materials.

Biogas plant waste

Ponni *et al.*, (2007) found that slurry which comes out of the biogas plant can be utilized by the farmers as manure for composting and vermicomposting.

Livestock waste

Livestock is mostly owned by all the farmers. Mehta *et al.*, (2002) have found that dung cakes could be made from livestock wastes. It could also be made into bio insecticide. Currently the energy consumption is rising and in today's pretext electricity generation is a big challenge. This raises the need of an alternate energy source. A study by Chaiprasert (2011) has shown the utilization of agriculture biomass to generate electricity.

Commercial Wastes

With the urbanization of cities more industries and automobiles have come in. This has resulted in large amounts of wastes every day. These can be from the markets, roads, buildings, hotels, commercial complexes, hostels, auto workshops or even printing press. The hospitals, nursing homes and medical institutes also release tremendous amounts of wastes which release toxins into the environment. Gaur *et al.*, (2015) studied the food industry waste and their use in nutraceuticals.

Disposable items from these units also produce many harmful chemicals causing danger to human health and infectious diseases. Other wastes which comes from mines and radioactive substances may cause a lot of damage to the society and environment.

Mining

The wastes include the over burden material, mine tailings (the waste left after ore has been extracted from rock), harmful gases are released by blasting (Collins *et al.*,1979). Zengxiang (2012) studied ways to dispose solid wastes from mines in transition from open pit to underground mining. Wu *et al.*, (2001) looked into the status quo and suggestions on comprehensively utilizing the mining tailings at home and abroad. Zhang *et al.*, (2005) surveyed the current situation and prospects of the comprehensive utilization of mining tailings.

Solid wastes in large quantities are obtained from mining during mineral processing activities which are dumped causing harm to the environment. These could be utilized for better soils (Zengxiang, 2012).

When the nuclear reactors maintain a measurable amount of radioactive waste material then it goes into the environment which is obtained from the mining of radioactive substances and from atomic explosion. Buyuksonmez *et al.*, (2000) studied the occurrence, degradation and fate of pesticides during composting and composting systems. Smith *et al.*, (2001) studied the effects of composts produced by a simple composting procedure on the yields of Swiss chard (*Beta vulgaris* L.) and common bean (*Phaseolus vulgaris* L.).

The mine wastes should be carefully utilized and the contamination from waste rocks should be controlled which needs to be environment friendly. Sand can be recovered from ore dressing from construction industries and by mine back filling approach on the guidance of disposing and utilizing the solid wastes from mines. Wu *et al.*, (2001) studied the status quo on the utilization of mining tailings at home and abroad. It was Xiao *et al.*, (2010) who studied the utilization of iron tailings at home and abroad. Deng and Yachao (2011) studied the hot-pressing molding of flagstone with iron-tailings powder. In a study by Zhang *et al.*, (2005) they were able to produce glass ceramics by iron tailings. Chen *et al.*, (2012) studied the tailings of vanadium for high performance ceramsite synthesis. Sirkeci *et al.*, (2006) studied the method to recover Co, Ni and Cu from the tailings of divrigi iron ore concentrator. Zhang *et al.*, (2005) studied the current situation and prospect of the comprehensive utilization of mining tailings. Maiti *et al.*, (2005) studied the method of metal accumulation by naturally growing herbaceous and tree species in iron ore tailings. In a study by Zhou (2007), practice on mining land rehabilitation and ecological reconstruction in China has been studied. Das *et al.*, (2000) explored the probability of iron ore tailing for the development of ceramic tiles. Zengxiang and Meifeng (2012) studied the methods for environmental restoration of a mine in transition from open pit and also underground mining. Xing *et al.*, (2001) studied the preparation of gold tailings from glass ceramics. Wang and Chang, (1999) studied the present conditions of China's iron tailings resource and its utilization.

Bio-medical wastes

Biomedical waste usually originates from hospitals or laboratories and mostly contains infectious materials. This could originate from packaging materials, unused bandages or even infusion kits. It has been seen that the wastes from research laboratories contains biomolecules or organisms which have disastrous effect in the environment. The blades or other discarded sharps and the gauges contaminated with blood are also categorized under biomedical waste. These wastes are highly infectious. Wastes from pharmacies include expired drugs and chemical wastes from testing laboratories which are highly hazardous (*Current World Environment, An International Research Journal of Environmental Science, ISSN: 0973-4929*).

Managing these wastes properly is important for direct positive health impact on the community. The health care workers may require specific treatment and management prior to its final disposal. Treatment methods like incineration have been found to be very effective. Land filling may cause wastage of large area of land when the contaminants leach out into ashes. In this way, the ground water also gets contaminated. Beri *et al.*, (1995) studied the nitrogen and phosphorus transformations as affected by crop residue management practices and their influence on crop yield. Beri *et al.*, (1992) studied the parameters to balance the nutrients and soil properties that can be affected by management of crop residues.

All different wastes can be segregated for effective disposal. The medical waste like sharps could be separated from the general municipal waste. This also prevents the illegal reuse of medical wastes which include syringes, needles and plastic materials. Some components could be recycled after thorough disinfection. Recycled plastic material can be used for non-food grade applications. Biodegradable waste can be composted to be used in garden. This can in turn enhance the revenue generation.

Biodegradable Wastes

Biodegradable waste are organic matter which may break down into carbon dioxide, water, methane or other organic molecules with the help of micro-organisms by using compost, aerobic digestion or anaerobic digestion. These are usually generated in food processing units, cotton mills, paper mills, sugar mills, textile factories and sewage. Frutos *et al.*, (2008) have discussed the strategies on ecologically sound fertilization for field vegetable production. Wastes from the slaughter houses are also biodegradable and some part of it is utilized. The skin is used in making shoes. However, when the quantity of these wastes become more then it acts as pollutants which can not be easily decomposed. Goyal *et al.*, (2009) studied the effect of rice straw compost on soil microbiological properties and yield of rice. Rao *et al.* (1996) have written an excellent article on compost science and utilization.

These wastes are divided into green and brown wastes. Scraps, fruits, vegetables and eggshells come under green waste and the others as brown waste. Meat and fish products should not be saved with the compost pile as they attract rodents and pests. If the collection bin is kept under the kitchen sink, the microorganism's breakdown the green and brown waste into compost. It has been seen that it is important to keep the pile moist, warm and well aerated. If the water is more the organic materials may rot instead of making compost. Temperature in a compost pile can be monitored with a thermometer which should be around 135° to 160° F along with enough air. Oxygen can be added just by turning the compost every two-four weeks (Jenkins, 1999). Use of rechargeable batteries is better as they are non-biodegradable wastes which can be recycled. Golueke (1991) has written well on the art and sciences of composting. Ministry of environment forests have notified biomedical waste rules (1998). Trillas *et al.*, (2006) studied the compost from agricultural wastes in cucumber seedlings using microorganisms.

Non-Biodegradable Wastes

These wastes can not be decomposed by microbes and are oxidized and dissociated. From colliery products like coal stone, metal scraps and sludge are generated, refineries produce dry inert solids and varieties of sludge containing oil. From thermal power plants ash is produced as solid waste. As these can not be reused they accumulate in the ecosystem. Somego into biogeochemical cycles. Non-biodegradable wastes also include DDT, pesticides, lead, plastics and mercuric salt as they cannot be decomposed by biological processes. These can be recyclable wastes with economic values which can be reused in plastic, paper, old cloth etc. Non-recyclable wastes do not have economic value of recovery such as paper, thermocoal and tetra packs. Disposal of non-biodegradable waste is a major concern which includes not just plastic but many other varieties of waste. Farrow *et al.*, (1969) studied the canning operations that could help in reducing the insecticide levels in prepared foods. Felsot and Dzantor (1990) worked on the ways to enhance the biodegradation of pesticides in the environment. There are a few ways to help non-biodegradable waste management.

Hazardous wastes

Highly reactive and toxic chemicals, biologicals, explosives and radioactive wastes create a great danger not only to the humans but also to the plants and animal lives (Table-2). These are known as hazardous wastes as these are very toxic. These can cause a lot of harm to human health and to the environment. These are in the form of solids, liquids, sludge and gases. Production of chemicals like lead, mercury, cadmium, chromium, leather, pesticides, dye, rubber and effluents from different industries generate such wastes. There is a danger in situations of inadequate storage, transportation or disposal. With the advent of disposable materials there has been an increase in the waste. On an average 400gms of waste (garbage) per person is generated per day. The quantity of non-biodegradable wastes as plastics is also increasing day by day. Wastes being generated by the hospitals and nursing homes are of three types that include domestic wastes, hazardous wastes and infectious wastes. Zaidon *et al.*, (2007) studied the properties of particle board made from pre-treated particles of rubber wood, EFB and rubber wood–EFB blend.

Floodgate (1984) studied with an objective to know what happens to the petroleum in marine ecosystems. Frontera-Suau *et al.*, (2002) studied the aerobic biodegradation of hopanes and other biomarkers by using crude oil-degrading enrichment cultures. Ycung *et al.*, (1997) studied the biodegradation of petroleum hydrocarbons in soil as affected by heating and forced aeration.

Hospitals discard many chemicals and consumables, pack medicines, release of radioactive materials which are very hazardous. Infectious wastes are generated in the hospitals and nursing homes. These spread many diseases. Under the Biomedical Wastes (Management & Handling Rules, 1998) it is the responsibility of the hospital to dispose the biomedical waste safely. There is a regulatory body which checks the implementation of the rules. Today, the people have a lot of awareness for better health, hygiene and what could spread various infectious diseases. The syringes used for the purpose of injections have been replaced by disposable syringes in place of reusable glass syringes. Unfortunately, this has increased the waste generation. These have shown to block the sewer pipes creating unhygienic conditions.

- Consumption of plastics by animals affects their health to a large extent.
- If the disposal is not done properly it attracts pigs which cause dangerous disease like brain fever.
- In case of leak from the disposal site it may contaminate the ground water.

3. b. Waste Reduction

Different types of wastes could be reduced by knowing how to eliminate/ reduce/ reuse or recycle various materials. Initially the biggest component of the waste should be taken into account and then be implemented.

- Re-refined petroleum products are less harmful and should be recommended for use.

Hydraulic oils should be used which are made from rapeseed oil and are (100%) vegetable oil. They are biodegradable, non-toxic and can be utilized at high temperatures and high viscosity than other petroleum products. Food by-products can be used as feed for animals. It will be cheaper if farmers make a cooperative. Recycled corrugated cardboards are sturdier. Wastes from construction area and during demolition of buildings could be reused and should provide extra container capacity for recyclingas many human diseases are associated with solid waste as the number of rats and flies increase and become the carriers of other organisms responsible for several dreaded diseases. The flies carry many pathogenic microorganisms for many types of dysentery and diarrhoea. Around 70,000 flies are known to be produced in one cubic foot of garbage. Mostly the wastes are made of organic compounds along with a number of inorganic minerals and harmful matter. This contaminates the environment resulting in the degradation of land, polluting drinking water, destructing aquatic life, degrading ground and surface water used for irrigation and industries. All these cause soil, air and water pollution.

The marine organisms are highly affected due to waste disposal. The growth of the marine algae goes down. In case of oil spills many beaches and estuaries get spoiled and that causes damage to marine lives. With winds many herbicides and pesticides reach the oceans ultimately which contaminates the marine water and damages the forests as that of mangroves. Pollution caused by thermal and radioactives disturb the coastal ecosystems which spoil the lives and breeding of fish. These wastes need to be converted into organic fertilizers which may reduce the pathogenic microorganisms, the weeds and the offensive odour (Converting Food Waste into Organic Fertilizer). Sebastian and Chaleff (1987) studied the soybean mutants with increased tolerance for sulfonylurea herbicides. WISErg Corporation (2015) have looked into the probability of converting food waste into organic fertilizer. They can also be extracted by processing of recyclables or by recovering the materials and resources or convert them to other energy forms (Pfafflin and Ziegler, 2006). Recovery of the resources is very important environmentally which is cost effective by conserving natural resources (U.S. environmental protection agency, 2012). This reduces the amount of waste for disposal and saving the space in landfills (Kjeldsen, 2002). Tanzi *et al.*, (2012) studied the terpenes as green solvents for the extraction of oil from microalgae. Resource recovery (as opposed to waste management) suggests the use of LCA (life cycle analysis) that offers alternatives to waste management. For mixed MSW (municipal solid waste) a number of broad studies have indicated that administration, separation from the source and collection followed by reuse and recycling of the non-organic fraction, energy and compost/fertilizer production of the organic material via anaerobic digestion can be used which will be very helpful. Most of the times resource recycling is beneficial as many items have precious metals and when recycled can be used as components in circuit boards. Resource recycling of wood chippings can be used as packaging materials in the horticultural sector, use in creating paths, walk ways, or arena surfaces. Atiq Uz Zaman (2009) did a life cycle environmental assessment of municipal solid waste to energy technologies. Strom (1985) has identified thermophilic bacteria which can successfully make solid-waste compost. Tiquia and Tam (2000) studied various parameters to evaluate pig litter compost.

3. c. Waste Treatment

Waste-To-Energy (WTE) is a new concept where the waste can be converted into a useable form of energy. It was in 1885 when the first U.S. waste incinerator was built which continued since then. It is found in Europe such as Scandinavia but not very common in the USA.

The waste is collected in three bin systems - blue to collect recyclables, green for compostables and black for landfill bound materials. In incineration method of disposal the solid organic wastes are burnt and converted into residue and gaseous products. In this process the solid waste are reduced to 20-30 percent of that the original volume. As these are done at high temperatures it is known as "thermal treatment" converting the waste materials into heat, gas, steam and ash. Alvarenga *et al.*, (2008) evaluated the composts and liming materials in the phytostabilization of a mine soil using perennial rye grass. Martin *et al.*,

(2011) established the analytical methods for the determination of ten steroid hormones in animal waste manure and agricultural soil using inverse and integrated clean-up pressurized liquid extraction and gas chromatography-tandem mass spectrometry. Conrad (2002) studied to control microbial methane production in wetlands of rice fields. Dale *et al*., (2010) have reported that waterborne outbreaks of gastrointestinal disease in Australia are predominantly associated with recreational exposure.

3d. Waste Disposal

Recycling of organic wastes

Organic wastes like residues from the crops, dung and urine of animals, slaughter houses, human excreta, sewage, biomass of weeds, wastes from fruits and vegetables, household wastes, sugarcane trash, oilcakes, press mud and fly ash from thermal power plants can be recycled. If some material can not be taken for direct application they are utilized by composting and vermicompost. For sustainable agriculture there is a need to develop a farming system which is profitable. It should be able to conserve the natural resources, protect the environment, enhance health and safety. There are two types of farming systems proposed to ensure sustainability. These are low input sustainable agriculture (LISA) and organic farming.

Low Input Sustainable Agriculture (LISA)

External production inputs are minimally utilized in this system reducing the production costs and the risk of the farmer. It also helps in reducing the pollution on the surface. This enables healthy food without pesticide residues. It has been seen that the high input system usually fails due to lack of economical and environmental sustainability. The conditions of soil, organic matter, air and water play an important role towards the growth of the plant. Lata *et al*., (2008) studied the technique of composting agro residues using fungal inoculants. Lawson *et al*., (1995) studied the various effects of organic matter on the growth, nodulation and nitrogen fixation of soybean grown under acid and saline conditions. Paul (2001) have studied the soil microbiology, its ecology and biochemistry in different conditions. Thomsen (2002) characterized the humic materials of different origin with a multivariate approach for quantifying the latent properties of dissolved organic matter.

Organic Farming and Waste Recycling

For sustainable agriculture it is important to do organic farming. In this approach chemical fertilizer, pesticides or herbicides are not used or used minimally. Conventional agriculture with high organic manure application could tolerate the pest and disease attack and helps in better crop growth and more food. However, the organic recyclable wastes are produced which include crop residues, farm industrial waste and sewage wastes. These provide nutrients to the plant. Due to biodegradation theorganic carbon becomes deficient in nutrients. Therefore, utilization of organic residues in agriculture is essential. It also protects the soil from erosion. The residues of crops have a high ratio of C: N and also immobilization nutrients. This can help in composing with efficient microbial inoculants. Beaudin *et al*., (1999) identified the key factors which affect composting of a weathered hydrocarbon-contaminated soil. Hue and Liu (1995) studied how the stability of the compost can be predicted. Hue *et al*., (1994) studied to enhance the levels of plant available phosphorus in an ultisol using yard waste compost. Perner *et al*., (2007) studied the effect of arbuscular mycorrhizal colonization and two levels of compost supply on the nutrient uptake and flowering of *Pelargonium* plants. Strom (1998) evaluated the pesticide residues in yard trimmings compost.

1. Sugarcane Trash Compost: Trash from fresh sugarcane contains (0.36%) N. The C: Nratio is 122:1. However, after composting the nitrogen content increases in the trash to higher content of N (1.09) with reduced C: N ratio (20.1). Per hector availability of trash is about 6- 8 (over all country about 19–38) million tons). Prince *et al*., (2000) studied the effects of coir pith compost on the growth and quality of leaves of the mulberry plant *Morus alba*.
2. Bio-gas Slurry: Organic manures obtained from animal wastes have high nutrients and help in building up soil fertility. In India alone around 1002 million tons of dung and 658 million tons ofurine is produced. If these wastes are utilized properly it can give about 5.7 million tons of N P K. There are many advantages of urine biogas as fuel (gas) and fertilizer (slurry). Dry slurry contains around 1.8% N, 1.10 % P_2O_5 and 1.50% K_2O.
3. Vermicompost: The nutrient content in vermicompost has been found to be much higher than other sources. Vermicompost contains (1.60%) N,(5.04%) P_2O_5 and (0.80%) K_2O. The C: N ratio of vermicompost is much lower (1.6) than the FYM (1.30). This provides essential nutrients to the plant crops and can be safely applied.

Industrial Wastes: Distilleries and sugar factories produce a lot of waste. The spent wash and molasses from distilleries and press mud from sugar factories can be utilized for improvement of soil. It has been found that the addition of press mud improves the soil fertility. It has the ability to improve saline and sodic soils. Coir waste from coir industry is also a very good source to improve soil health. The waste is estimated approximately up to 12.5 t/ha. It has been proved as a good fertilizer for groundnut and increase in the yields of maize.

Municipal and Sewage Wastes: It is that with organic manure the yields of the crops can be enhanced without any side effects. Municipal and sewage waste plays an important role towards the contribution of organic manure. In India, the total municipal waste is about 12 million tons/annum which contains different minerals like (0.5 %) N, (0.3%) P and (0.3%) K. Sewage (liquid portion) sludge (solid portion) is available to an extent of 4 million tons /annum containing (3 %) N, (2 %) P and (0.3%) K. However, care needs to be taken as these might contain metals which are hazardous to plants, animals and also to human beings. Paré *et al.*, (1999) studied the extractability of trace metals during co-composting of biosolids and municipal solid wastes. Alibardi and Cossu (2015) studied the composition variability of the organic fraction of municipal solid waste and effects on hydrogen and methane production potentials. Paredes *et al.*, (1996) worked on the influence of the bulking agent on the degradation of olive-mill wastewater sludge during composting. Paredes *et al.*, (2005) studied the effect of olive mill wastewater in composting and its impact on the compost on a Swiss chard crop and soil properties. Perez-Murcia *et al.*, (2005) worked on the use of composted sewage sludge as horticultural growth media and studied the effects on germination and trace element extraction. Sukesan and Watwood (1998) studied the effects of hydrocarbon enrichment on trichloroethylene biodegradation and microbial populations in finished compost.

Crop residues: After harvesting of crops some residues are left. These are in the form of straw. Straws from different cereals and residues contain high amount of nitrogen, phosphorus and potassium being (0.5%) N, (0.6%) P and (1.5 %) K. The crop residues can be easily recycled in compost. Hussein *et al.*, (1996) used oil palm shells for the preparation and characterization of active carbons. Sueli *et al.*, (2002) studied the chemical composition of *Pleurotus pulmonarious* (Fr.) Quel as substrate and residue after cultivation. Birdie *et al.*, (2014) have studied the banana byproducts as underutilized renewable food biomass with great potential.

Rice Husk: Rice milling industries produce large amount of husk as waste. Although the nutrients are very low quality namely (0.3 %) N, (0.2 %) P and (0.3%) K but it can be added to the manure. The physical condition of the soil can be improved by adding rice husk into the wet soil. It is also good to use in saline and alkaline soils to improve the soil health. Rice husks are also utilized as a bedding material for many animals.

Biocultures: There are many microbial cultures which improve the soil condition and also help in higher yields of the crops. The generally used cultures are namely PSB (*Bacillus*, *pseudomonas*, *Aspergillus*, *Penicillium*, *mycorrhiza*) and many more.

Green Manuring Crops: *Sannhemp, Dhaicha, Glyricidia, Sesbania.*

If the wastes are converted into beneficial products or energy they help in many ways.

Benefits of Waste-To-Energy (WTE): As the conversion of waste to energy is most of the times a very economical process it is preferred. Although it is the location and other factors which really decide the cost of WTE. Table (3) gives the country based wastes and energy recovery in those countries. If the process is nearby placed as that of the waste then the cost of garbage can be avoided as one can save on transportation and landfilling and energy can be generated economically. Madhumithah *et al.*, (2011) studied the ways to utilize vegetable wastes for the production of protease by solid state fermentation using *Aspergillus niger.*

Conserves Natural Resources: In using WTE usage of natural resources as of oil, gas and coal are not utilized. This saves these resources for other uses. Other processes apart from WTE use these resources. As per literature a single WTE facility (Huntsville Waste to Energy facility) saves more than 200,000 barrels of oil per year.

Better Recovery of Materials: During the process of WTE it is also possible that the metals which are left in the waste stream can be extracted from the residue. This can also be recycled.

Less Impact on Climate: As WTE produces very less greenhouse gas it is beneficial to the environment. It is mainly methane gas that is released from the decomposition of the waste stream in landfills. It is very harmful much more than carbon dioxide. It is estimated that it is almost 20 times more harmful and is responsible for devastating climate change.

By the method of composting the waste could either be disposed or can be treated by controlled aerobic decomposition of organic waste materials. This happens with the help of small invertebrates and microorganisms. The most common composting techniques include static pile composting, vermicomposting, in-vessel composting and many others.In anaerobic digestion the organic materials are decomposed. It uses oxygen and bacteria-free environment to decompose the waste material where composting must have air to enable the growth of microbes. Dukare *et al.*, (2011) evaluated a novel microbe which could amend composts as biocontrol agents in tomato.Grohmann *et al.*, (1994 a,b) studied the fermentation of galacturonic acid and other sugars in orange peel hydrolysates by the ethanologenic strain of *Escherichia coli*. Grohmann and Baldwin (1992) studied the hydrolysis of orange peel with pectinase and cellulase enzymes. Grohmann and Bothast (1994 a) generated pectin rich residues by the processing of citrus fruits, apples and sugar beets. Grohmann *et al.*, (1995) studied sugar fermentation in orange peel hydrolysates to be converted to ethanol by recombinant *Escherichia coli* KO11. Renu Agrawal and Patwardhan (1993) studied the production of peroxidase enzyme by callus cultures of *citrus aurantifolia*. Renu Agrawal *et al.*, (1991) studied the formation of flavouring constituents in callus cultures of lime, *Citrus aurantifolia* S. Mamma and Christakopoulos (2008) have written an excellent review on successfully hydrolyzing both orange and grapefruit peel waste using commercial cellulase and pectinase enzymes to glucose, galactose, fructose, arabinose, xylose, rhamnose, and galacturonic acid. Singh *et al.*, (2012) studied the utilization of agro-industrial wastes for the simultaneous production of amylase and xylanase by thermophilic Actinomycetes. Allen *et al.*, (2001) studied a comparison between hot liquid water and steam fractionation of corn fiber. Gurpreet Dhillon and Surinder Kaur (2016) studied the agro industrial wastes as feedstock for enzyme production.The strategies were addressed to enhance value addition of agro-industrial wastes in order to develop sustainable technology. This aspect is very important in a bioprocess. It has been observed that the treatment of the residue from industrial process requires fewer efforts. Also value addition of these wastes can be done at a very low cost. In an integrated process the waste of one process after treatment can be used as raw material for another process mainly through biotechnological pathways. Before initiating any process many valid points need to be looked into. It is important to know the nature of the residue, which is the source of carbon and its type, which nutrients and/or what toxic compounds are present. At the beginning the pretreatment steps should be worked out so that the deterioration of the material can be avoided.It is also important that in the presence of all these metabolic changes many beneficial products can be formed by biotransformation. For the detoxification of the residues pretreatment may be a must containing anti-metabolites. Pretreatment may be from drying or concentration, grinding, improvement either by thermal or enzymatic treatment, balancing the nutrient content, reduction of the concentration of toxic compounds, to the transformation of recalcitrant compounds. These may require different aspects of technical and economical aspects. These steps directly or indirectly affect the complete process. Silicon compounds are essential for many products in day today use. Large amounts of these are produced every year. Some of the useful qualities include polishing of the silicon wafers which are utilized in electronics, strengthening tyres, making paints and paper white, as an abrasive in toothpaste, many tiny packets of silica gel beads put in various products to absorb moisture. For the production of pure silicon compounds a very expensive procedure is required. This involves heating mined silicon metal and anthracite coal to 3,500 degrees celsius in an electric arc furnace. Michigan University have innovated a new technology where pure silica can be produced from agricultural waste. The researcher who developed the process says it could save approximately six tons of carbon emissions per ton of silica compounds produced. He estimates the cost of the technique to be 90 percent less than the current process with virtually no carbon footprint. This was developed by a professor named Richard Laine, of materials science and engineering. The new technique is believed to be the first simple, inexpensive chemical method for producing high-purity silica compounds from agricultural waste. Much of the world's agricultural waste contains silica, and the search for a practical way to extract it stretches back 80 years. While the new process could be used to produce silica and silicon-containing chemicals from many types of agricultural waste, Laine focused on using the hulls left over from processing rice. The hull is the outermost layer of the rice grain. It is removed when rice is processed. Out of tons of husks produced per year many are burnt to produce electricity and the left over ash contains high levels of silica. This ash is used in construction or as insulation, but much of it is dumped in landfills. A major challenge is to remove the strong chemical bond between silicon and oxygen. This is supposed to be one of the strongest that exists in nature.Salciccioli *et al.*, (2011) found two easy and inexpensive ways to break the bond present in ethylene glycol or antifreeze by differentiation of O–H and C–H bond scission mechanisms of ethylene glycol on Pt and Ni/Pt using theory and isotopic labeling experiments. The antifreeze combined with a small amount of sodium hydroxide weakens the chemical bonds between the silica and the rice hull ash at the beginning of the process, dissolving the silica into a liquid solution.The solution is then heated to 390 degrees celsius forming a polymer of silica and antifreeze. The process requires energy. When the hulls are burned the carbon is released which was infact absorbed by the rice plant otherwise. Because of this the process is known to be carbon-neutral. Due to heating silica-antifreeze polymer is produced which is removed after filtration. At the end of the process grain alcohol is added. If seen chemically it resembles antifreeze and can replace antifreeze easily and can also be recycled. Pure silica products can be prepared after distillation of liquid silica. Laine has formed a Michigan company named Mayasil, to commercialize the technology for manufacturing process. According to Laine this will bring a fundamental change in the way silica products are made and used.This will allow producing high-purity silica and other silicon compounds right next to the rice fields. It will be possible to process rice and produce high-grade silica in a single location with little or no carbon footprint. Laine recently received the 2015 Michigan Green Chemistry Governor's Award from the Michigan Department of Environmental Quality for this work. Mayasil is a spinoff of Mayaterials, a company Laine founded in 2003 in conjunction with U-M Tech Transfer. Mayasil holds a patent on the silica production process.A paper (Laine *et al.*, 2016) on the ways to avoid carbothermal reduction on distillation of alkoxysilanes from biogenic, green and sustainable sources is very informative. Andary *et al.*, (1985) studied a caffeic glycoside, poliumoside, ester from teucrium belion. Andary *et al.*, (1982) also studied the structure of caffeic acid sugar esters verbascoside and orobanchoside obtained from Orobanche rapum-genistae. Hameed *et al.*, (2009) prepared an oil palm empty fruit bunch-based on activated carbon for the removal of 2, 4, 6-trichlorophenol by using response surface methodology. Ermawati *et al.*, (2007) studied ways to use cow manure waste by degrading and studying the behavior of its natural steroid hormones during biological treatments and ozone oxidation. Satish *et al.*, (2015) were able to extract anthocyanins from winery waste material. Kimber (1973) studied the phytotoxicity from plant residues. They also studied the relative effect of toxins and nitrogen immobilisation on the germination and growth of wheat. Lua and Guo (2001) prepared and characterized activated carbons from oil-palm stones by gas-phase adsorption method.

Plastic Waste: If we look into our daily lives plastic has become indispensable. Plastic products are very common. According to a survey, the production of plastic water bottles per annum utilizes almost 1.6 million barrels of oil.The disposal time for plastics is very long as it does not decompose. Normally, plastic items can take up to 1000 years to decompose in landfills. But plastic bags take 10-1000 years to decompose while plastic bottles may take 450 years or more.

Disposable Diapers: In America alone more than 18 billion disposable diapers are thrown away every year. In landfills it takes almost 250-500 years to decompose. This is causing a great havoc to the environment.

Aluminum Cans: In America alone more than 120,000 aluminum cans are recycled. For complete decomposition of Aluminum cans around 80-200 years are required in landfills.

Glass: As glass is made of sand it is much easier to recycle it. It is also possible to make new glass by breaking the old one and melting it. However, once glass thrown into thelandfills; it may take millions of years to decompose. Some scientists have also reported that many a times it doesn't decompose at all in the landfill.

Paper Waste: Paper occupies the largest space in American landfills. The time taken by papers to decompose is 2-6 weeks in landfills. However, it has been worked out that recycling paper items, can save landfill space, while reducing the energy and virgin material requirements of making non-recycled paper.

Food Waste: On the basis of weight, food waste occupies the largest area of American landfills. The time taken for food waste decomposition depends on the type of food. Normally, an orange peel takes 6 months but an apple core or a banana peel takes around one month to decompose. An important component of food recycling is having the right container to deal with it. LeBlanc (2015) studied the Yuck Factor and the First 100 yards of food waste recycling. Different materials take different time to decompose in the landfills. Time taken by different major product watses to decompose is cigarette butts take 10-12 years; monofilament fishing line takes 600 years; rubber boot sole take 50-80 years; foamed plastic cups take 50 years; leather shoes take 25-40 years; milk cartons take 5 years; Plywood takes 1-3 years; painted board takes 13 years; cotton gloves take 3 months; cardboard takes 2 months; Styrofoam does not biodegrade at all; nylonfabric takes 30-40 years. As per the deconstruction institute, every year United States building deconstruction and demolition industry produces around 124,670,000 tons of debris that goes into the landfills and only 2 percent of total waste building materials are recycled. Maria *et al.*, (2012) studied the antioxidant activity and consumer acceptance of grape seed flour-containing food products.

Thermal Treatment: In thermal treatment of waste a lot of heat is utilized. There are various techniques to use it like in cineration is very common method of thermal treatment. The waste material is burnt in the presence of oxygen. By this the waste is converted into ash, gas, water vapor and carbon dioxide. By this method energy can be recovered and can be utilized in the form of electricity or for heating purpose. There are many advantages to this approach as waste volume is reduced at a fast pace, the transportation costs are much less and also the harmful greenhouse gas emissions are much less. Other two methods are gasification and pyrolysis in which organic waste materials are decomposed by exposing waste to low amounts of oxygen and very high temperature. In the method of pyrolysis oxygen is not utilized whereas in gasification a very low amount of oxygen is required. As in the method of gasification energy is obtained without any air pollution it is preferred over pyrolysis. Open burning have no pollution control devices and release harmful substances like hexachlorobenzene, dioxins, carbon monoxide, particulate matter, volatile organic compounds, polycyclic aromatic compounds and ash. However, this method is generally seen as it is an inexpensive way to get rid of solid waste.

Ashoush and Gadallah (2011) studied the large amount of wastes of mango juice industry in the form of mango peels and seed kernels. These have high levels of natural bioactive compounds which help in the prevention of diseases. According to the studies it has been found that mango peel powder (MPP) has high levels of ash, crude fiber and water holding capacity. Mango kernel powder (MKP) has been estimated to have higher fat and protein than MPP. Scientists have prepared biscuits using both peel powders and have found improvement in antioxidant properties and the nutritional quality was found to improve. Mango seed kernels have been found to contain a considerable amount of total phenolic compounds, total lipids, unsaponifiable matter and a low amount of crude protein. Mango seeds obtained from Nigeria contains 10.06% crude protein, 14.80% oil, 2.62% ash, 2.40% crude fibre, 70.12 % carbohydrate and energy content 453.92 KJ/100 g. Mango seed is very rich in glutamate (13.00 g/100 g of protein) while methionine has the lowest value (1.04g/100 g of protein). Mango (*Mangifera indica* L.) belongs to the 45% to 75% of the whole seed (Beyene and Araya 2015). Soong *et al.*, (2013) studied large amounts in millions of tons of mango wastes which were available with high phenolic contents. They are a good source of phytosterols as campesterol, β-sitosterol, stigmasterol and tocopherols. These are utilized in many products such as puree, nectar, leather and pickles. Ravani *et al.*, (2013) have studied the phenolic compounds in Egyptian canned mango slices. These are high in tannins, gallic acid, coumarin, ellagic acid, vanillin and mangiferin being the major compounds (Chakraborthy *et al.*, 1991). Some mango varieties have high concentrations of ferulic acid, cinammic acid and some unknown compounds. Schiber *et al.*, (2003) and (NunezSelles, 2005) referred that the thousand tons of mango fruits were produced in 2009. Mango peel has been the major contributor for polyphenols and sesquiterpenoids. Peel contributes about 15-20% of the fruit NunezSelles (2005). Renu Agrawal *et al.*, (2003) studied the biotransformation of ferulic acid to vanillin by locally isolated bacterial cultures. Nankai *et al.*, (1998) studied the biotransformation of (±)-Lavandulol by the Plant Pathogenic Fungus *Glomerella cingulate*. The use of wastes for production of value added products such as improved feed, bio-processing reagents with selective catalysts, safe green chemicals, bio-fuels, biogas, bio-plastics and biopolymers would serve the agro-processing sector more resource efficient and sustainable. This will be vital for its competitiveness and will support the rural livelihoods through increased demand for local crops and bio-resources, enhancing the agribusiness opportunities for farmers and small scale enterprises in the region. Promoting the conversion of waste into renewable energy (such as biogas, bio-fuel etc) will also reduce the need to import costly fossil fuel and serve as a step to mitigate climate change. Promotion of innovation on local small-scale bio-refineries at the village level will add value to their local crop produce.

3. e. Clean Technology

The definition of waste management involves all the activities that are required to manage waste from the source to its final disposal destination (Dikeman,

2009). The different steps involve the collection of waste, transportation, treatment and disposal of waste with continuous monitoring. Each country has the legal and regulation formalities to follow to dispose the waste. The authorities have also set up rules for recycling. Wastes, could be from the raw materials, obtained during the processing of raw materials into intermediate and final products. The consumption of final products or other human activities include municipal (residential, institutional, commercial), agricultural, and social (health care, household hazardous wastes, sewage sludge especially in India. (Anjello and Ranawana, (1996). The basic idea of waste management is to reduce the harmful effects of waste for health promotion and to improve the environment. Duah-Yentumi and Kuwatsuka (1980) investigated the effect of organic matter and chemical fertilizers on the degradation of benthiocarb and MCPA herbicides in soil. Marín *et al.*, (2006) studied the bioremediation by composting heavy oil refinery sludge in semiarid conditions. Meysami and Baheri (2003) screened the fungi and bulking agents for contaminated soil bioremediation. Ciafardini and Zullo (1998) have studied the inhibitory activity of oil mill waste water. Klinkhamer and Klean industries (2009) have forcasted the three main clean technology sectors.

According to Diaz (2011) the practices to utilize wastes are different in various countries (developed and developing nations), regions (urban and rural area) and sectors (residential and industrial). In the clean technology there is improvement in energy efficiency. The use of resources should be sustainable and must protect the environment. These include issues related to recycling, renewable energy (wind power, solar power, biomass, hydropower, biofuels, etc.), information technology, green transportation, electric motors, green chemistry, lighting, greywater and many others. Environmental finance is a method by which new clean technology projects that has proven that they are "additional" or "beyond business as usual" can obtain financing through the generation of carbon credits. A project that is developed with concern for climate change mitigation (such as a Kyoto Clean Development Mechanism project) is also known as a carbon project. While there is no standard definition of "clean technology," it has been described by Clean Edge, a clean technology research firm, as "a diverse range of products, services and processes that harness renewable materials and energy sources, dramatically reduce the use of natural resources, and cut or eliminate emissions and wastes." It notes that "Clean technologies are competitive with, if not superior to, their conventional counterparts. Many also offer significant additional benefits, notably their ability to improve the lives of those in both developed and developing countries". By 2018 there will be three main clean technology sectors namely, solar photovoltaics, wind power and biofuels which may bring revenues upto $325.1bn (Dickinson *et al.*, 2011). A Nationwide Survey was done by Pandey *et al.*, (2009) on the availability of Indian biomass resources for exploitation. Srokol *et al.*, (2004) studied the hydrothermal upgrading of biomass to biofuel in which they studied on some monosaccharide model compounds. Sundar *et al.*, (2014) used electron beam pretreatment of switchgrass to enhance enzymatic hydrolysis to produce sugars for biofuels. Kumar, *et al.*, (2009) worked on the methods for pretreatment of lignocellulosic biomass for efficient hydrolysis and biofuel production. Rati Rao *et al.*, (2003) worked on the fermentation technologies for smaller communities. According to Ionel ioana (2010) waste is an important environmental pollution source for the soil and also for ground water. It also pollutes the air. As per the European standards in all EU countries waste deposit is not allowed in open land fields. According to the national regulations all the exiting non-ecological deposits need to be closed and have to be turned into ecological ones. A thorough evaluation of waste management is required. In fact it is an energy source which needs to be protected and not to be wasted. Turning the waste into good business involves reactions like combustion, fermentation and recycling giving rise to cleaner energies. There are many barriers which need to be corrected at the level of society and by policy makers. Utilization of fruit wastes in producing single cell protein has been studied earlier (Mondol, 2012). Effects of rapeseed residue on lead and cadmium availability and uptake by rice plants in heavy metal contaminated paddy soil was well studied by Ok *et al.*, (2011). Ciafardini and Zullo (2000) have studied alpha glucosidase activity in olive brine during the microbiological debittering process. It was Nascimento *et al.*, (2015) when they produced and characterized a new fibronolytic protease from *Mucor subtillissimus* UCP 1262 in solid state fermentation. USDA (1992) is a field handbook which deals with agricultural waste management.

4

Useful Products from Waste Utilization (Reuse/ Recovery/Recycling)

All these methods of utilization and the conversion to useful energies depend on the climate, population density of the area and the basic infrastructure. It is important to work out the market for the products which are produced.

Bio-waste management: It needs to take from small actions to the ambitious ones. European Commission proposes to use wastes which could make up to 25% of total energy consumption by 2020. They propose to have directives for the promotion of renewable electricity (Directive 2001/77/EC) and biofuels (Directive 2003/30/EC). The waste production in Europe has gone upto 2 kg per capita per day. Recycling rates are also rising (European Commission, 2017). The Member states must develop a national action plan and outline the national policies to develop existing biomass resources and mobilise new biomass resources for different uses. A European project has studied the amount of renewable energy to be used in 2020. According to them it will be around 195 million tons of oil equivalents (Mton) of biomass which is biodegradable part of MSW. This will be 20% of renewable energy target. It has also been estimated that the amount of bio-energy from the MSW will be 20 Mton, which would account for around 7% of all renewable energy in 2020. This may be available after composting and anaerobic digestion (Atanasiu, 2010).

Paper mill wastes: It is well known that the liquid waste from the paper mills is fermented to produce Torula yeasts, using *Candida utilis*. In Pekilo process spent sulphite liquor consists of monosaccharides, polysaccharides and carbohydrate derivatives. Spent sulphites and acetic acid are utilized in the production of waste liquor from sugar and potato industries. The process produces protein, sulphite and other material burnt for energy. By this method it is possible to produce about 10-15 thousand tons of protein annually from 1OQ thousand ton waste. Vijayendra *et al*., (2003) worked on the screening and isolation of gel forming exopolysaccharide producing microorganisms from soil samples.

Meat waste: There are vast livestocks in India. It is growing at a very fast rate of 6% per annum. It is mainly the poultry and fish which have pushed up the GDP of the country to more than 40% in total agricultural sector. This could go up further in case the animal by-products could be utilized efficiently. It will improve the economy and reduce environmental pollution in the country. If the utilization is not proper a lot of potential revenues are lost which will lead to enhanced cost of disposing these products. Non-utilization of animal by-products in a proper way may create major aesthetic and catastrophic health problems (Jayathilakan *et al.*, 2012).

In India there are a large variety of livestocks which are growing at the rate of 6% per annum. Poultry and fish contribute substantially in enhancing the GDP in India. This accounts for >40% in the agricultural sector. If the by-products are utilized efficiently it helps the economy and prevents the environmental pollution in the country. On the other hand, if the by products are not utilized properly or are under utilized it may lead to loss of potential revenues. The cost to dispose these waste products may also go very high. If the animal by-products are not utilized in a proper way it may cause many health problems. Processing wastes from meat, poultry and fish industries can be recycled and can also be converted into high valued beneficial products. Looking into the safety aspects a number of countries restrict the use of meat by products. Apart from blood, many organs like liver, lung, kidney, brain and spleen have very high nutrition. They are also important medicinally and have pharmaceutical importance. Many applications hold for treated fish waste which is an important animal feed. They are also used for biodiesel/biogas, dietectic products (chitosan), natural pigments (after extraction) and cosmetics (collagen).

4a. Biofuel

By the use of contemporary biological processes it is possible to get biofuel. It could be from agriculture or by anaerobic digestion. It is not produced from resources like coal and petroleum or any other prehistoric biological matter. Reactions like transesterification and esterification are involved for the production. Vegetable or animal fats and oils react with short-chain alcohols to produce biodiesel. Ethanol is usually utilized as it has low molecular weight and has low cost. Bioconversions using methanol give higher yield. The reaction is catalyzed by using either acid or a base. However, commonly base-catalyzed transesterification reactions are used. Here the catalyst costs are much lower than by acid catalysis. However high sensitivity is observed both towards water and free fatty acids present in the oils (Chai *et al.*, 2014).

Commonly the plants used are yellow grease (recycled vegetable oil), pure / "virgin" vegetable oil and tallow. If in case recycled oil is used it is processed to remove impurities from cooking, storage, and handling, such as dirt, charred food, and water. Virgin oils also need to be refined. However the purity may not be upto the food-grade level. Degumming to remove phospholipids and other plant matter is common, though refinement processes vary (Münch, 2007). Base-catalyzed transesterification removes water from any type of the feedstock. In the process the triglycerides are hydrolyzed forming salts of the fatty acids (soaps) instead of producing biodiesel. Atiq (2009) has made a life cycle environmental assessment of municipal solid waste to energy technologies.

Transesterification

Esters are formed by using animal and plant fats and oils. These are composed of triglycerides formed by the reactions of three free fatty acids and the trihydric alcohol: glycerol. In this process the added alcohol (methanol or ethanol) gets deprotonated with a base to make it a stronger nucleophile. The reaction has triglyceride and the alcohol. Most of the times this reaction is very slow or does not start at all. Therefore, to initiate the reaction either heat or some catalysts (acid and/or base) are used. They act just as catalysts as they do not react in the reaction. Common catalysts for transesterification include sodium hydroxide, potassium hydroxide and sodium methoxide. The virgin oil conversion is the most economical process. It requires very low temperatures and pressures with a conversion rate of over 98%. Bioconversion by other methods may require acid catalysis which is usually very slow (Fei *et al.*, 2014) as it is the predominant method for commercial-scale production.

Ultrasonic reactor method: In this method the ultrasonic waves cause the reaction mixture to produce and collapse the bubbles constantly. It provides both the mixing and heating which is required simultaneously to carry out the transesterification process. Thus, using an ultrasonic reactor for biodiesel production drastically reduces the reaction time, reaction temperatures and energy input. Industrial scale ultrasonic devices allow for the higher scale processing of many thousands of barrels per day.

Bio-Ethanol: One of the most common biofuel is bioethanol which is around 90% of the total biofuel usage. Conventionally, it is an enzymatic conversion where in starchy biomass gets converted into sugars which on fermentation forms ethanol of fuel grade. This bio fuel is produced from a number of feedstocks, many different crops which include cereal crops, corn (maize), sugar cane, sugar beet, potato, sorghum and cassava. Many coproducts are formed as animal feed which make the process economically beneficial. From sugar cane the conversion into sugar is easier. After crushing the stalk called as bagasse it gives heat and power for energy applications. The biggest plants in the world are situated in Brazil for bio-ethanol (sugar-cane ethanol) and in the

United States (corn ethanol). Sometimes this fuel is mixed with gasoline in different blends of 5%-10%. Blending is done in flex-fuel vehicles. In Brazil, a minimum of 22% bioethanol must be present in gasoline. In recent times the focus is to utilise ligno-cellulosic materials. The use of different wastes include cellulosic wastes, maize stover, cereal straw, food processing wastes and switch-grass. These can be obtained from integrated crops. Ultimately, it helps in increasing land availability. In this process the biomass is pre-treated to release cellulose and hemicellulose. Then it is hydrolysed to release fermentable 5- and 6-carbon sugars. The sugars are then fermented and separated from the solid residues and ultimately distilled to obtain fuel. A lot of research is underway for better conversions. Hydrolysis is usually inhibited by lignin and other components. They can be extracted to be used as fuel which may reduce the cost and emissions.

Cost –The cost of biofuel is always variable. It depends on the raw material, the process being followed, the cost of the land and labour costs. The market for the byproducts, agricultural subsidies, food (sugar) and oilmarket needs to be seen. Ethanol energy content by volume is two-third as that of gasoline, therefore the costs refer to litre of gasoline equivalent (lge). According to the literature survey the price of sugar cane ethanol in Brazil is \$0.30/lge which is free-on-board (FOB). This price is quite competitive to gasoline at oil prices of \$40-\$50/bbl (\$0. 3-\$0. 4/lge). It has been found that in other regions the costs are usually more than \$0. 40-\$0. 50/lge. However, the prices of ethanol from other sources like maize, sugar-beet and wheat seem to be less which is around \$0.6-\$0.8/lge (excl. subsidies) which could be further reduced to \$0.4-\$0.6/lge. If the prices of ligno-cellulosic ethanol are seen it costs around \$1. 0/lge at the pilot scale, if a basic feedstock price is \$3.6/GJ for delivered straw. However, the ethanol production from cereals may cost \$10-\$20/GJ. The costs can be reduced if there is improvement in the process, in the scaling up procedures, the costs of the waste feedstocks are reduced along with co-production of other by-products (bio-refineries). If the costs for biodiesel are compared to biodiesel from the animal fat it seems to be the best option as it is cheapest (\$0.4-\$0.5/lde) and therefore economically viable. If trans-esterification is done for the vegetable oil it will cost around \$0.6-\$0.8/lde. The approximate cost of BTL diesel which is obtained from ligno-cellulose is around \$0.9/lde (feedstock \$3.6/GJ).

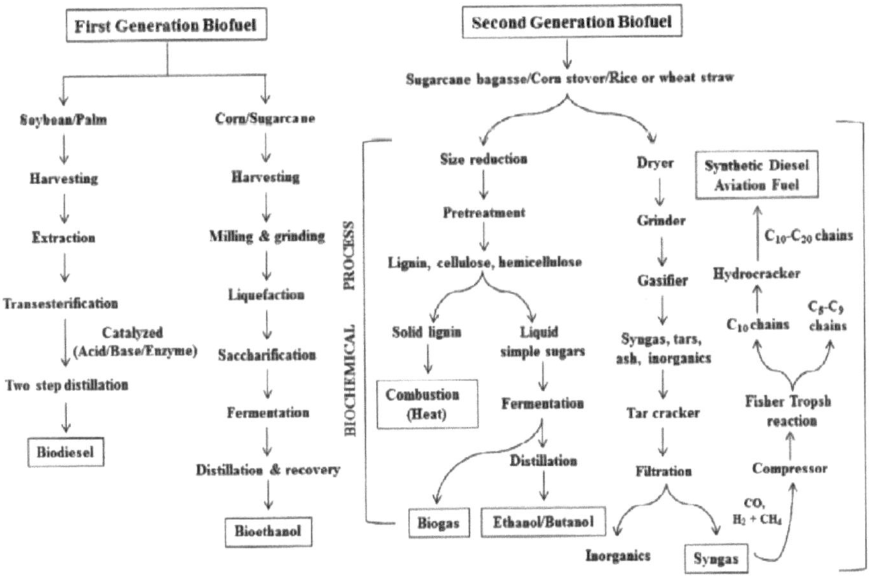

Fig. 1: Biocatalysed-production of fuel ethanol from lignocellulosic biomass
Source: (Panel Juliet Ben-Iwo, *et al*., 2016)

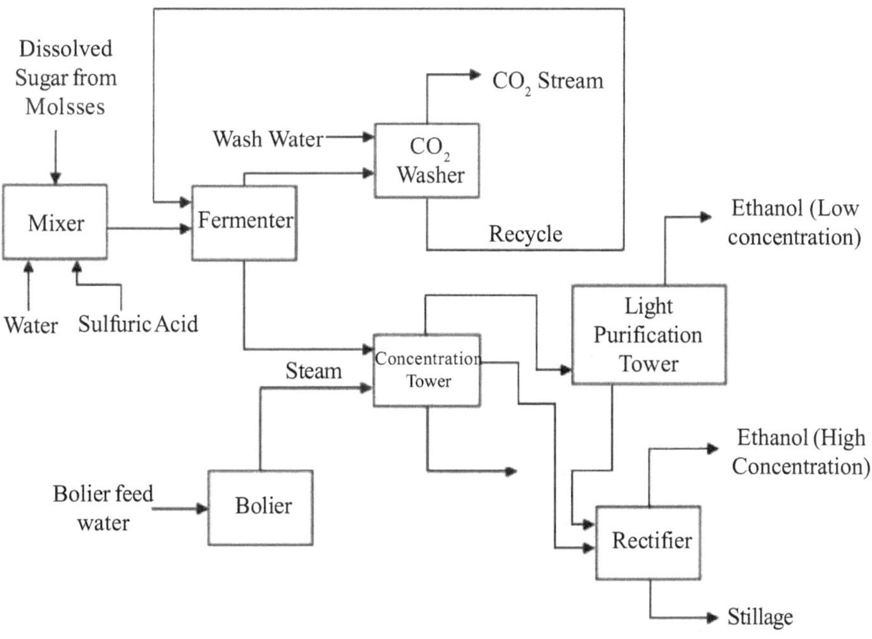

Fig. 2: Flow diagram to obtain Ethanol from Molasses.
Source: Ruhul Amin M., Saquib Hossain M., Sarker M(2013)

Scientists have worked on five strains of yeast which can turn agricultural by-products namely; straw, sawdust and corncobs into bioethanol which is a well known alcohol based biofuel. Each year it contributes to more than 400 billion litres of bioethanol from crop wastage. This is because high temperatures and acid conditions are necessary in the glucose release process. However, in this process the waste is broken down into compounds which are toxic to yeast (furfural and hydroxymethyl furfural). It is noticed that they make the fermentation process difficult. According to researcher Dr. Tom Clarke, from UEA's School of Biological Sciences "Bioethanol is a very attractive biofuel to the automotive industry as it mixes well with petrol and can be used in lower concentration blends in vehicles with no modifications". Vehicles which run on bioethanol have been on the roads since 1979 in Brazil. Agricultural residues can be utilized by a number of conversion technologies.

Conversion of wheat straw or corn stover to cellulosic ethanol can be done via pretreatment, hydrolysis and fermentation (Fig. 1, 2). However, wheat straw and other agricultural wastes to BtL through thermochemical pathways are very common and are effective. It is also possible to convert mixed agricultural wastes into bio-crude by thermochemcial conversion. Innovative conversion pathways based on catalysis and biotechnology can also be used to convert agricultural and other cellulosic wastes to biofuels.

Agricultural Wastes of the Palm Oil Industry by Hydrothermal Treatment

Recently, a lot of attention is being given for the utilization of agricultural waste of oil palm trunks. The conversion to solid biofuel was done using hydrothermal treatment. Presently, this waste is used as an organic fertilizer in plantation fields. The slow release of CO_2 and CH_4 gas on decomposition gives bad smell, which has been found to last up to a year. In 2005, approximately 43. 05 million tons of fronts and 13. 94 million ton of trunk wastes were generated in Indonesia, the largest crude palm oil producer in the world. This amount will continue to increase with the rapid growth in the Indonesian palm oil industry. Thus, a better method to manage such wastes is highly desired. Various treatment methods for several agricultural wastes have been reviewed (Arvanitoyannis, 2007, 2008 a,b, c). For energy application purposes, the methods can be simply divided into thermo-chemical and biological processes (Warnecke, 2000). Hydrothermal treatment is one of thermo-chemical process which has attracted much attention in recent times. In this process, biomass is treated with hot compressed water which yields in gases, aqueous organics and upgraded solid (McKendry, 2002, Ni *et al.*, 2006). After filtration and drying the obtained solids can be used for solid fuel due to its high calorific value. The temperature in hydrothermal treatment is much lower (200-350°C) as compared to 450-550°C for pyrolysis and 900–1200°C for gasification (Zhang *et al.*, 2010). In addition, biomass conversion takes place in a wet environment and therefore, high moisture content of feed biomass is not a problem. In this treatment, water works as a medium and also as a chemical reactant on decomposition. This makes the method suitable for treating biomass with high moisture content, like agricultural wastes which contains more than 50 wt. % of moisture in fresh condition. It is different in pyrolysis and gasification where there is a limitation of moisture content in the feed (Zhang *et al.*, 2010; Kruse, 2009).

Scott *et al.*, (1949) have discussed about the Banana by-products which are an under-utilized renewable food biomass with great potential. Among the many fruit crops Banana (Musaceae) is one of the world's most important one that is widely cultivated in tropical countries. It is valued for its applications in the food industry. The by-products are good source for other industries by recycling the agricultural waste. It does not allow the loss of huge amounts of biomass and helps environmental issues. Utilization of banana by-products can be done in various food and non-food applications. These include peels, leaves, pseudostem, stalk and inflorescence. These serve as thickening agent, coloring and as flavors. This also serves as an alternative source for macro and micronutrients, nutraceuticals, livestock feed, natural fibers and sources of natural bioactive compounds and bio-fertilizers. These should be turned into high valued commercial by-products for sustainability. This will also provide additional income to small scale farming industries with better quality and safe products in the market. Scott *et al.*, (1949) studied the partial purification and properties of antibiotic substances from the banana (*Musa sapientum*). Gassaraa *et al.*, (2016) screened various agro-industrial wastes to produce ligninolytic enzymes by *Phanerochaete chrysosporium*. Ponnuswamy *et al.*, (2016) studied the bioconversion of agro-industrial wastes for the production of fibrinolytic enzyme from *Bacillus halodurans* IND18.

4b. Enzymes

Many industries utilize enzymes for various purposes. Recently there is a lot of development in enzyme technology especially from microbes. They provide many benefits as they are economical, give high yields batch after batch, give good consistency to the product, down stream processing is easy and the technology can be easily optimized. As it does not depend on seasonal fluctuations the supply of product is mostly regular. As the doubling time of microbes is on cheap media the processes are very economical to be persued. These are utilized in medical field which involves diagnosis, treatment, biochemical investigation and monitoring many dreaded diseases.

Kumar and Duhan (2011) have isolated and characterized amylase enzyme from *A. niger* MTCC-104 by using solid state fermentation technique. Mahalakshmi and Jayalakshmi (2016) studied amylase, cellulose and xylanase production from a novel bacterial isolate *Achromobacter xylosoxidans* isolated from marine environment. Singh *et al.*, (2010) studied the utilization of agro xylanase by thermophilic actinomycetes. Suganthi *et al.*, (2011) studied the production of amylase using agri waste in *A. niger*.

Commonly two enzymes namely amylase and lipase have been studied extensively which have great importance in different industries and therapeutic industry. These can also be used for various purposes like bioremediation, agriculture and pharmaceuticals. Ismail (1996) investigated the production of multi enzyme preparations containing pectinase, cellulase and xylanase using six fungal isolates. In their study they used *Aspergillus niger* 2, *A. niger* A-20, *A. oryzae* 1911, *Memmoniella* sp. 6, *Penicillium chrysogenum* 3486 and *P. oxalicum* 7. All the strains were grown on waste orange peels which were utilized as a carbon source. Among all the strains *A. niger* A-20 was found to be the most potent. It produced highly active multienzyme systems. Renu *et al.*, (2010) utilized agro-industrial wastes for the simultaneous production of amylase and xylanase by using thermophilic actinomycetes. Seyis and Aksoz (2005) investigated the use of apple pomace, orange pomace, orange peel, lemon pomace, lemon peel, pear peel, banana peel, melon peel, and hazelnut shell as substrate for xylanase production using *Trichoderma harzianum*. Withmelon peel as the substrate in SSF the activity was found to be maximum. Other wastes including apple pomace and hazelnut shell also had good activity. *Sporotrichum thermophile* produces a thermostable polygalacturonase under submerged culture and citrus peel as carbon source (Kaur *et al.*, 2004). The waste utilization includes a major part of citrus peels utilization for the production of enzymes mainly pectinolytic ones. These enzymes hydrolyze the pectic substances. Pectinases act in three ways. It can be either in pectin, pectic acid or oligo-D-galacturonate forms. The most preferred substrate is galacturonate form. The action reaction for pectinases can be either by transelimination or hydrolysis. The cleavage could be random (endo, liquefying of depolymerizing enzymes) or endwise (exo- or saccharifying enzymes). The three major types of pectinases are pectinesterases, depolymerising enzymes (enzymes that either hydrolyze glycosidic linkages or cleave 1,4-glycosidic linkages by transelimination) and protopectinase (Jayani *et al.*, 2005). Pectinolytic enzymes are of significant importance (Singh *et al.*, 2012). Gassaraa *et al.*, (2016) studied and screened the agro-industrial wastes to produce ligninolytic enzymes by *Phanerochaete chrysosporium*. Vijayaraghavana *et al.*, (2016) studied the bioconversion of agro-industrial wastes for the production of fibrinolytic enzyme from *Bacillus halodurans* IND18. Dhillon and Kaur (2016) studied the agro-industrial wastes as feedstock for enzyme production, their exploitation and usefulness. Bodie *et al.*, (1994) studied the economically important organic acid and enzyme products from *Aspergillus* fungi.

Brar *et al.*, (2015) have published a good book which deals with the enzymes in value-addition of wastes. Bok *et al.*, (2001) have shown the derivatives of Flavanone and its composition for preventing or treating blood lipid level-related diseases by using wastes.

It is important to know the basic needs of the resource from the view of supply and demand. The biomass which needs to be exploited is usually very heterogeneous as the origin could be either from grown crops or from crop residues of food and feed production. It could be a residue from the forest or even from the flora under water. Other potential wastes include municipal waste, manure, animal products and industrial wastes. These are used for bio based products.Value added products from different biomasses are obtained with the use of enzymes. Looking into this, enzymes play a very pivotal role in producing these products.

The uses of enzymes play a very major role for industries in producing energy, organic chemicals, polymers, fabrics and health care products. In general, an enzyme based bio-economy offers many benefits and opportunities:

- Innovation in economic growth and development for the biomass resources
- Innovative business spin-offs
- Better energy security as independent of non-renewable resources
- Innovation in linking the agricultural and industrial sectors. This will better the business pursuits and environment.
- Innovation in promotion of sustainability
- Greenhouse gas emissions will be reduced
- Innovation in improving health security as exposure to harmful substances will decrease through the utilization of enzymes instead of hazardous chemicals
- Employment generation and economy development.

As it can be noted that there are a number of opportunities and benefits using enzymes. There are many natural and potentially cheap and abundant polymer which can be used as agricultural waste (wheat straw, corn stalks, soybean residues, sugar cane bagasse), industrial waste (pulp and paper industry), forestry residues, municipal solid waste, etc. (Wiselogel *et al.*, 1996). Lignocelluloses create around 50% of the biomass in the world which is 10–50 billion tons per annum (Classen *et al.*1999). Several researchers (Widmer *et al.*, 2010) have successfully hydrolyzed both orange and grapefruit peel waste using commercial cellulase and pectinase enzymes to glucose, galactose, fructose, arabinose, xylose, rhamnose, and galacturonic acid. According to them glucose, fructose and galactose from hydrolyzed citrus peel waste can be fermented to ethanol by *Saccharomyces cerevisiae* yeast. From pectin hydrolysis galacturonic acid, arabinose and xylose as well as other sugars are fermented by *Escherichia coli* K011 producing ethanol and acetic acid. *E. coli* KO11is a recombinant bacterial strain. This has been developed to ferment arabinose, xylose and other hexoses to ethanol (Beall *et al.*, 1991). Orange peel oil concentration must be reduced in the hydrolysate before the fermentation (Widmer *et al.*, 2010). When yeast is grown on either orange peel oil and/orD-limonene (a monoterpene that makes up more than 90% of orange and grapefruit peel oils) then the growth of yeast is seen to slow down. This effect has been observed by many earlier workers (Agricultural research service, 1956). The mechanisms by which limonene and other monoterpenes similar in structure to limonene inhibit yeast function and growth have been studied by many workers (Brennan *et al.*, 2001). Wilkins *et al.*, (2007) reported that whey is used as a cheap fermentation medium for ethanol production from *X.campestris* which utilizes lactose as sugar (Papoutsopoulou *et al.*, 1994). Citrus by-products can be used as ruminant feeds (Bampidis and Robinson (2006). Beall *et al.*, (1991) studied ethanol production from xylose and other sugars by recombinant *Escherichia coli*. Beg *et al.*, (2001) studied the microbial xylanases and their industrial applications. Bilanovic *et al.* (1994) studied the production of xanthan from citrus waste. Blanco *et al.,* (1999) have studied the production of pectic enzymes in yeasts. Bocco *et al.*, (1998) studied the anti-oxidant activity and phenolic composition in citrus peel and seed extracts. Cardona and Sánchez (2007) studied fuel ethanol production. Carrol *et al.*, (1999) studied how the citrus limonoids, flavonoids as well as tocotrienols can be utilized for the treatment of cancer. Chau and Huang (2003) studied the chemical composition and physicochemical properties of different fibers prepared from the peel of *Citrus sinensis* L. Claasen *et al.*, (1999) studied as to how the biomass can be utilized for the supply of energy carriers. Danieli *et al.*, (1990) studied regioselective acylations of flavonoid disaccharide monoglycerides which were enzyme-mediated. de Gregorio *et al.*, (2002) studied the fermentation of lemon pulps for SCP and crude pectinase production. In a report de Vries and Faubion (1999, 2003) defined dietary fiber. Renu Agrawal and Patwardhan (1987) studied the production of thymol from callus cultures of *Carum copticum* L. Renu Agrawal *et al.*, (1990) studied the different parameters affecting allicin formation as a secondary metabolite by static cultures of *Allium sativum*. Dhakal *et al.*, (2005) worked on the wastes of potatoes, pumpkin, cauliflower, cabbage and brinjal to study the production of protease by solid state fermentation using *Aspergillus niger*. Estimation of specific activities for individual substrates conferred the highest upon pumpkin with 13.44 umg protein/ml, with comparable results obtained for cauliflower and cabbage. This study presents a novel economical approach for the bioconversion of vegetable wastes for the production of protease that is industrially significant.de la Fuente *et al.*, (2011) have worked on the impact of fresh and composted solid olive husk and their water-soluble fractions on soil heavy metal fractionation, microbial biomass and how the plant uptakes them.

Single cell Proteins

Malys *et al.*, (2011) investigated the production of proteins in *Saccaharomyces cerevisiae*. Using cucumber and orange peels they found that higher amount of crude protein (53.4%) was obtained from cucumber peels. It is predicted that as cucumber peels contain higher amount of available carbohydrates and minerals that would have affected favorably in the production of yeast biomass. It was also observed by the scientists that even orange peel contained high concentration of carbohydrates but less mineral content (3.55%) than that of in cucumber peels (6.96%) and was unable to enhance biomass production (Mondol, 2012).

India is the largest mango producing country in the world. This constitutes around 40% of the world's mango production (12.3 MT). Globally the mangoes produced were 30.7 million tons in 2010 with a growth rate of 2.7%. Other major mango producing countries include China, Thailand, Mexico, Pakistan, Philippines, Indonesia, Brazil, Nigeria and Egypt. All of them are tropical countries. Processed mango products are the major goods for export. In that case there are several millions of tons of mango seed wastes which are produced annually from factories. Seeds constitute around 13% portion of a mango. Statistics shows that this constitutes around 1.67 million tons of waste in India alone which has a high potential nutraceutical and has a lot of food value.

Soil Fauna Responses: Better livelihoods can be given to the rural folks through increased demand for local crops and bio-resources. This will also increase the business opportunities for agricultural farmers and motivate many small scale enterprises to come into the region. When conversion of waste will be promoted into renewable energy (biogas, biofuel) then it will reduce the need to import costly fossil fuel. This might help in improving the climate change. If innovation can be promoted at village level and assistance can be provided to the local people informing local small-scale bio-refineries then it will add value to their local crop produce.

Anaerobic digestion is a process and the best way to deal with bio waste. Land fill may not be the right place for the organic waste. There are also problems with leachate and gas. The technologists had found ways to convert organic waste into compost and fertilizers. This method is considered the most innovative in the developed countries. It gives a large-scale solution to the organic waste and also permits the resulting gases to be converted into energy. The wastes get degraded by microorganisms in the absence of oxygen. It can be used to treat organic solid waste and wastewater also. It is a rapid method and the remains can be used as fertilizer while the biogas produced gets converted into energy. It is therefore a source of renewable energy. Waste to energy (WTE) has many advantages. Recently, newer types of WTE have come into the systems which are much in use. These include gasification, pyrolysis, thermal depolymerization and plasma arc gasification.

Gasification and plasma arc gasification are helpful in converting organic materials into synthetic gas (syngas). This is made up of carbon monoxide and hydrogen. The gas is then burnt to produce electricity and steam. A plasma gasification plant uses a plasma torch that works at similar temperature as that of sun surface. This creates an environment in which solid or liquid waste turns into syngas. It breaks down the molecular bonds present in the waste. It gets converted to energy so that the waste disappears completely.

Zero Waste is another approach in which the products and processes are managed systematically. In this method it eliminates the volume and toxicity of waste and materials by conserving all resources. There is no burning and the resource can be recovered.

The International Solid Waste Association (ISWA) had established a task force in November 2007 to study the interaction between waste management and the production of green house gases. Proven, practical and cost effective technologies are being tried by the waste sector which can contribute to GHG mitigation and helps in securing significant global GHG emission savings. Waste prevention, minimization, reuse and recycling are increasing around the globe.

Recently all around the world biowaste is utilized along with the traditional agricultural practices. This mainly includes addition of manure and slurry to the soil. This also helps in recycling of nutrients and improves the quality of organic matter. Today, the amount of animal waste available is much larger in amount (Animal Waste, Water Quality and Human Health. Ed. Al Dufour *et al.*, 2012). Lately as the agricultural and livestock breeding activities have increased the number of livestock it has also increased the production and accumulation of large amount of wastes. Along with this the use of mineral fertilisers and pesticides has increased to get more fodder production resulting in less agricultural production. Therefore, more awareness for the addition of organic waste to the soil has become significant. A problem to the environment is another major issue especially of atmospheric pollution which needs to be tackled.

4. c. Flavonoids

There are many phenolic compounds in grapes (*Vitis vinifera* L.) which are an important source of antioxidants (Andjelkovic *et al.*, 2013).The dietary polyphenols have higher antioxidant activity than vitamins (Banini *et al.*, 2006 and Anastasiadi *et al.*, 2010). Wine and wine by-products have become important for their high antioxidant and antimicrobial properties and their potentially beneficial effects for human health (Sun *et al.*, 2002; Baydar *et al.*, 2006). The benefits of phenolic compounds have been worked out well (Lee *et al.*, 2005; Banini *et al.*, 2006 and Del Pozo-Insfran *et al.*, 2007). The world consumption of wine is high. The nutrition is imparted by polyphenolic compounds like flavonoids, anthocyanins and tannins. These compounds affect the sensory properties of wines which include color, flavor, bitterness, astringency and aroma (Gomez *et al.*, 2012). It is also important to check the quality of grapes for wine making (Caceres *et al.*, 2012). While selecting grape varieties for wine production one needs to check the economic yield, resistance to pests and diseases, juice recovery, sugar, acid content and the functional properties like anthocyanins, phenolics and tannins for good wine quality. Teixeira *et al.*, (2014) have written a good review on the natural bioactive compounds from winery by-products to be used as health promoters.

In India various varieties of wine grapes are being utilized to produce wines. Most of them grow under tropical conditions (Adsule *et al.*, 2012). Around 51 varieties are grown in Indian tropical conditions. To obtain better quality of grapes many methods are being established which include pruning and single cropping cultivation practices (Jogaiah *et al.*, 2013). Karibasappa and Adsule 2008 and Adsule *et al.*, 2012) have worked out for the best grape variety based on juice yield and must quality parameters. These include the Shiraz, Cabernet Sauvignon, Merlot and Zinfandel among red varieties and Sauvignon Blanc, Ugni Blanc, Vermentino and Garganega among white varieties are recommended for commercial wine production. Sharma *et al.*, (2012) also revealed the varietal impact on phenolic content and antioxidant activities of wines. The phenolic profiling of grapes is equally very important. Scientists have worked on the profiling of phenolics in wine grapes. This helps in knowing the phenolic composition at harvest stage and to establish a base for differentiation of wine grape varieties based on the phenolic composition as well as the phenolic maturity of grapes. Rudra *et al.*, (2015) have noted that 18 per cent of the fruit and vegetable production which amounts to Rs. 44,000 crores goes as waste annually in India alone. The processing of fruits and vegetable is very low which accounts for 4% only. It has been estimated that 20% of fruits and vegetables are exported to other nations. In India it is mostly catered in-house to the defence, industrial sectors and commonly for household consumption. For proper utilization of wastes it is important that fruit and vegetable processing industries are centralized in India. This is found to be difficult as a large number of units are at the cottage level or home scale and small scale levels. These also have limited capacities with maximum being 250 tons/annum. However, the big Indian and multinational companies in the sector have large capacities in the range of 30 tons per hour or so. Waste by definition means anything unused or not used to its full advantage. In household wastes more of rotten fruits and vegetables are found. However, industry wastes contain less of over or under-ripe fruit or vegetable, rotten organic matter but more of cellulosic waste. These include peels and seeds. The use of this composite mass of cellulosic or fibrous waste along with carbohydrates gives rise to amines of phytochemicals. Most of the vegetables and some of the fruits contain around 25%-30% of this type of waste. Depending on the plant species, variety and tissue, high levels of health protecting antioxidants, such as vitamin C and E, phenolic compounds including phenyl propanoids and flavonoids and or carotenoids such as lycopene have been found. Being a good source of phytochemicals and antioxidants these wastes can be utilized effectively. The entire tissue of fruits and vegetables are rich in bioactive compounds, which include phenolic compounds, carotenoids and vitamins. In many cases the activity of antioxidants and antimicrobials have been found to be higher in waste products than in the final produce. The new aspects concerning the use of these wastes as by-products have gained increasing interest on the production of food additives or supplements with high nutritional value as they are high valued products with easy and economical down stream processing. There are many by-products which can be formed from these wastes. These include sugars, minerals, organic acids, dietary fibre and phenolics which have a wide range of action. The activity of these compounds could be anti-tumoral, antiviral, antibacterial, cardio protective and ant imutagenic activities. Many workers have found during their investigation that the percentage of phytochemical compounds is more in the peel and the seeds as compared to the edible tissue. Even the phenolic content and flavonoid content in the by-products were found to be higher than the final products. This has been proved by many scientists especially when they have experimented with mango seeds and peels. Gorinstein *et al.*, (2001) found that the total phenolic compounds in the peels of lemons, oranges and grape fruits were 15% higher than that of the fruit pulp. Peels of other fruits were also tried and it was found that apples, peaches, pears, yellow and white flesh nectarines had double the amount of total phenolic compounds than that in the fruit pulp. Phenolic compounds are very high in apple peels which contain up to 3300 mg/100 g on dry weight basis. Edible pulpy portion of bananas (*Musa paradisiaca*) also have high content of phenols approximating 232 mg/100 g of dry weight. According to Someya *et al.*, (2002) this amount is about 25% of that present in the peel. Pomegranate peels contain 249.4 mg/g of phenolic compounds as compared to only 24.4 mg/g phenolic compounds found in the pulp of pomegranates. This trend has been found in many other fruit seeds like mangoes, longans, avocadoes and also in jackfruits. In a study on grapes on the manufacture of grape juice and white wine it was found that the seeds and the outer skin are high source of many phenolic compounds. These include mono, oligo and polymericpro-anthocyanidins. George *et al.*, (2004) while working with tomatoes found that the peels and seeds of tomatoes had high phenolic content than the pulp of the tomatoes.The peel byproduct of tomato has significantly higher levels of total phenolic-compounds, total flavonoids, lycopene, ascorbic acid and antioxidant activity as compared to the pulp and the seeds. Corrales *et al.*, (2008) have worked on the extraction of anthocyanins from grape by-products assisted by ultrasonics, high hydrostatic pressure or pulsed electric fields which will be beneficial. Samman *et al.*, (1998) studied the dietary aspects of flavonoids and coronary heart disease. Arnold (1931) had worked on the agricultural waste of the industries. He found that red grape pomace (RGP) and onion solid wastes (OSW) can be used as raw material to produce flavonoid-enriched extracts. This could be done using ultrasound-assisted solid-liquid extraction. The extraction medium used was composed of water and glycerol. The extraction of flavonoids from both materials was shown to obey first-order kinetics. Maximum diffusivities (De) values were 4.01ˆ10′11 and 2.35ˆ10′11 m2 s′1, for RGP and OSW extraction, respectively, while the corresponding activation energies (Ea) were 14.00 and 15.23 kJ¨ mol′1.

It has been analysed that annually millions of tons of agricultural wastes in the form of cornstalks, corncobs, straws and hulls are produced. This waste is made up of cellulose, lignins and pentosans in high proportion. After proper segregation this waste can be used in many different uses especially for wood. Insulating board and paper are made from cornstalks and straw. Pressed board and Maizolith have been tried for the production from cornstalks. Corncobs have been found useful for the formation of furfural, charcoal, acetic acid, methanol, acetone, tar, oxalic acid and xylose. Oat hulls are also good source of furfural and are also utilized in the formation of many other industrial products. Cornstalks and sewage have been used in the production of illuminating gas.

According to Balasundrama *et al.*, (2006) people have become very aware on the importance of phenolic compounds as these compounds have high antioxidant activity and are able to quench free radicals. They are helpful in inhibiting many diseases. These are found in almost all plants and play a very important role in the welfare of human beings. The main feature of these compounds is the presence of an aromatic ring which has one or more hydroxyl group. Structurally, it may be simple phenolic molecule or a complex high-molecular weight polymer. Flavonoids, constitute the major phenolic group in plants. These have C_6–C_3–C_6 structure. There are more than eight thousand different phenolic compounds and more than half of the total phenolics have this structure. The antioxidant activity of any phenolic compound depends on the structure, number and the position of the hydroxyl group and the nature of substitutions on the aromatic rings. In the human diet especially for vegetarians the major sources of phenolic compounds are fruits, vegetables and beverages. The processing industries make various food products and agricultural products generate large quantities of by-products which are rich in phenolics and have high antioxidant activity. Many scientists have studied the effect of these byproducts and have proved that these compounds are effective sources of phenolic antioxidants. These have also been tested in edible oils and animal products from fish, meat and poultry. Scientists have found these phenolic rich extracts to impart high antioxidant activities. Vijayendra *et al.*, (2015) studied the effects of probiotic cultures on fermentation of Moringa leaves extract.

These phenolic compounds have also shown to have antiallergic, antiartherogenic, anti inflammatory, antimicrobial, anti thrombotic, cardio protective and vasodilatory properties (Benavente-Garcia *et al.*,1997; Manach *et al.* 2005). Puupponen-Pimia *et al.*, (2001) studied the antimicrobial properties of phenolic compounds from berries.

Health benefits from consuming high levels of fruits and vegetables are attributed to phenolic compounds (Hertog *et al.*,1993; Parr and Bolwell, 2000). These are due to their antioxidant activity (Heim *et al.*, 2002). Phenolic compounds could be a major determinant of antioxidant potentials of foods (Parr and Bolwell, 2000) and could therefore be a natural source of antioxidants.

Survase (2012) used different agricultural wastes to grow *Pleurotus sajor* caju (Fr.) on wastes like wheat straw, paddy straw, soybean straw, pigeon pea straw and green gram straw. He also used different combinations like wheat + soybean, wheat + pigeon pea, wheat+ green gram (1:1) and wheat alone. Various parameters like fresh weight, fruiting bodies and protein content were measured. As analysed *Pleurotus sajor* caju has high protein content of 20-35% (dry weight basis). This is much higher than many other vegetable and fruit. Reducing pollution by utilizing agricultural wastes is very beneficial. Indirectly it helps the farmers.

In a study different lignocellulosic agro wastes were used, for the cultivation of, oyster mushroom. The highest protein percentage was found when it was cultivated. In India, paddy and wheat straw have been used for commercial production of oyster mushroom (Jain and Vyas, 2002). Paddy straw was found to be the best substrate for cultivation of oyster mushroom (Bano and Shrivastava, 1962; Bonatti *et al.*, 2004).

Ligno-cellulosic materials can be converted into protein rich food by non-conventional method with the use of fungi. According to various studies (Bhatti *et al.*, 1987; Thampi *et al.*, 1966; Bonatti *et al.*, 2004) wheat straw was found to be the next best substrate after paddy straw to cultivate oyster mushroom. Bahukhandi and Munjal (1989) have tried Sorghum straw. Bhandari *et al.*, (1991) tried *Echinochloa frumentacea* (Poaceae), *Eleusine coracana* (Poaceae) and grasses to grow *Pleurotus sajor* caju. Patrabansh and Madan (1997) have studied the chemical constituents of the mushroom.

In the citrus food industry the wastes generated after citrus juice extraction constitutes about 50% of fresh fruit weight. The solid residue consists of the peel (flavedo and albedo), pulp (juice sac residue), rag (membranes and cores) and the seeds. A major problem is the disposal of fresh peels for the processing industries. Most of the times, these industries dry the residue in the solid form which is sold as a raw material for pectin extraction or in the form of a pellet which is taken for animal feeding. Due to low protein content and high amount of sugar it is not found good as animal feed. The application of agro industrial byproducts in bioprocesses offers a wide range of alternative substrates, thus helping to solve pollution problems which are related to their disposal. However, recently a lot of research has been undertaken to use citrus byproducts for generating value-added products like enzymes, single cell protein, natural antioxidants, ethanol, organic acids, polysaccharides and prebiotics. Workers have used citrus peels as a substrate for biotechnological applications. Over the past few decades, an increasing trend towar defficient utilization of natural resources has been observed around the globe. Disposing of agro industrial waste results in loss of biomass. Bioconversion of this waste into different metabolites into high valued products is very beneficial and has high commercial value (Thomsen 2005). Production of enzymes has been worked out by many scientists using citrus peel (Larios *et al.*, 1989; Garzón and Hours 1992; Ismail 1996; Silva *et al.*, 2002; Dhillon *et al.*, 2004, Mamma *et al.*, 2008). Other products tried are bioethanol (Grohman *et al.*, 1994b, 1998; Wilkins *et al.*, 2007), citric acid (Aravantinos-Zafiris *et al.*, 1994; Rivas *et al.*, 2008), xanthan gum (Bilanovic *et al.*, 1994; Gales *et al.*,1988; Stredansky and Conti,1999) single cell protein (de Gregorio *et al.*, 2002), prebiotics (Olano-Martin *et al.*, 2000; Hotchkiss *et al.*, 2003; Premsuda *et al.*, 2012) and natural antioxidants (Li *et al.*, 2006; Man-dalari *et al.*, 2006) among many others. There are many products and by-products from different parts of tissues from citrus fruits (Bambidis and Robinson, 2006). Whole peel or rind (pericarp) consists of exterior yellow known as flavedo or epicarp and interior white called as albedo, the spongy peel is known as mesocarp. It has been found that albedo is rich in pectin. The whole peel along with molasses is a good feed for animals. It is also used in human foods and food supplements. However, the pulp which is known as endocarp is mainly used to produce raw juice for human nutrition using different methods. Cores, segment walls or membranes, juice vesicles and seeds consist the total peel residue. The pulp residue combined with peel residue is used as feeds. From the lime treated mass peel and pulp residues, citrus processors produce byproducts such as press liquor, citrus molasses, citrus pulp, citrus meal and feed yeast. Utilization for human foods and food supplements is also common.

Seeds

These are used to produce seed oils, seed meals and cake after pressing dried seed. Activated sludge and yeasts can also be produced.

Biofilms

From tomato processing industry a lot of waste is obtained in the form of polysaccharides which is very useful. Granadilla peels are utilized in the development of biodegradable films by means of biopolymers and glycerol solutions (Strazzullo *et al.*, 2003; Tommonaro *et al.*, 2007). From these polysaccharides, films are obtained which are solid, clear and elastic. These biofilms have biotechnological applications in different fields like agriculture. In the case of polysaccharides extracted from lemon pomace no results were obtained using the same procedure for biofilm production.

In recent times the consumer is aware of the food composition and human health. Therefore, the consumption of fruits and vegetables has increased. Consumer demand for foods that are enriched in bioactive compounds and nutraceuticals have also increased.

Reuse of agro-industrial wastes which are rich in healthy plant ingredients have been given a lot of attention. Work on grapes has shown that these are high in bioactive compounds. Statistical analysis has shown that 210 million tons of grapes (*Vitis vinifera* L.) are produced per year. Out of this around 15% grapes are used in wine making industry. This gives rise to solid waste (approximately 30%, w/w of the material used). The wastes obtained from wine industry are solids which are biodegradable. These include stems, skins and seeds. In the winery by-products many bioactive compounds were found which have health promoting activities. These were studied and proved by both *in vitro* and *in vivo* studies and have also shown beneficial effects on human health (Teixeira *et al.*, 2014).

Food processing industries produce a large amount of waste. Utilization of this waste is practically very difficult. At the same time it is also very challenging all around the world. The formation of this waste takes place whenever the finished product is produced. Many studies have shown the generation of phytochemicals, antioxidants, dietary fibre, food ingredients which include pectin, natural colour, vitamins, antibiotics, proteases, ethanol and biogas from various wastes (Rudra *et al.*, 2015). The wastes are from different sources like fruits, vegetables, cereals, milling of pulses and coffee processing units.

The processing of fruits and vegetables is only around 4% of the total. The export of fruits and vegetables is estimated to be about 20%. In India the processed foods are mainly for defence, institutional sectors and for household consumption. As these processing industries in India are not centralized the capacity of production is small. These are either at the cottage level or are established on home scale level. The capacity per year is found to be up to 250 tons. However, the big multinational companies in the sector have large capacities which are in the range of 30 tons per hour. If profit needs to be accounted from waste treatment then it is important to segregate waste from a single fruit or vegetable. Later it can be treated in order to produce different types of pharmaceutical products and nutraceutical products. They may also be utilized as food ingredients. These days the farmers are forming clusters and this approach has shown many advantages.

The phenolic compounds present in pomegranate peels (249.4 mg/g) are much higher as compared to the pulp (24.4 mg/g). Total phenolic compounds in the seeds of mangoes, longans, avocadoes and jackfruits were higher than that of the edible product. Gorinstein *et al.*, (2001) found that the total phenolic compounds in the peels of lemons, oranges and grapefruits were 15% higher than that of the pulp of these fruits. Total phenolic compounds in the peels from apples, peaches, pears as well as yellow and white flesh nectarines are found to have doubled the amount than that of the fruit pulp. Apple peels were found to contain up to 3300 mg/100 g of dry weight of phenolic compounds. Grape seeds and skins, the byproducts of grape juice and white wine production, apart from many phenolic compounds contain mono, oligo and polymeric proanthocyanidins (Shrikhande, 2000).

Similar results were found in tomatoes. The peels and seeds of tomatoes have been found to be richer sources of phenolic compounds than the pulp of the tomatoes. Total phenolic compounds, flavonoids, lycopene, ascorbic acid and antioxidant activity were higher in the peel of tomato than the pulp and seeds (George *et al.*, 2004).

4. d. Anthocyanins, Flavones and Flavonols

The world's most important fruit crop is Banana (*Musaceae*). It is mostly cultivated in tropical countries which have many applications in many food industries. They are very good source of highly valuable raw materials for other industries which are obtained by recycling of the agricultural waste. In this way loss of biomass can be prevented. Scientists have worked extensively to utilize banana by-products which include peels, leaves, pseudostem, stalk and inflorescence in various food and non-food applications. They can be used as thickening agent, coloring agent, flavor, alternative source for macro and micronutrients, nutraceuticals, livestock feed, natural fibres and sources of natural bioactive compounds and bio-fertilizers. Most important is to know the sustainability of utilization and how much is the feasibility of these by-products. The by-products obtained should have the ability to be converted into expensive commercial outputs.This will make the process sustainable. Small scale farming industries will also have additional source of income with good quality and safety. Miraje *et al*., (2015) studied the extraction of anthocyanins from winery wastes.

Antioxidant Properties

In recent times many consumers use sunflower oil for cooking purposes. In this respect the stability of the oil is important. Scientists have found that addition of 400 ppm methanol extract and 5% mango seed kernel oil could increase the oxidative stability of sunflower oil when it is incubated at room temperature. Similar results were found even during frying.

Quality characteristics of fresh and stored potato chips were studied and it was found that both extract and oil improved the stability. These results provide useful information on the utilization of mango seed kernel as natural antioxidant in foods. Mango seed kernels obtained from Thailand have very high amounts of phenolic antioxidants, metal chelators and tyrosinase inhibitors. 1, 2, 3, 4, 6-Penta-O-galloyl-beta-D-glucopyranose, methyl gallate and gallic acid have been identified as components of an ethanolic extract of Thai mango seed kernel of cultivar Fahlun. However, it was noticed that in Nigerian mango variety which is commonly known as Ikanekpo, the level of tannins (45 g kg^{-1}) were very high. An amount of 48% could be extracted in a procedure where soaking and thermal treatment are combined when flour is formed. When it was subjected to chromatographic techniques around six phenolic compounds and eight phospholipids were found. Phenolics and phospholipids isolated from mango seed kernel, when added jointly to buffalo ghee, helps in extending the shelf-life of ghee. In order to reduce colour loss and lipid oxidation, the mango seed extract has been found to be very useful (0.1 or 0.2%). It has been seen that the seed extracts inhibit growth both that of gram negative and gram positive bacteria. Opportunistic gram negative bacteria, *Ps. aeruginosa* (ATCC 27853) was found to be the most sensitive strain. Hepato protective properties of the mango kernel extract (MKP) have been reported by determining their *in vitro* anti-inflammatory activities and by evaluating their hepato protective potential against liver injury in rats induced by carbon tetrachloride (CCl_4). Mango seed kernel extract has been found to have high total phenol content, radical scavenging activity, metal-chelating and tyrosinase inhibitory activities. This extract is found to be suitable in food, cosmetic, nutraceutical and pharmaceutical products. Therefore the applications are huge and beneficial. Either the extract or the oil was found to improve the stability and quality characters of fresh and stored potato chips. These results provide useful information on the utilization of mango seed kernel as natural antioxidant in foods.

Colors

In recent times the demand for eco-friendly/non-toxic colorants is growing. Importantly, it is found that eco friendly colors improve health (Miraje *et al.* 2015). Obtaining colors from natural sources is very important and is an emerging area of science in today's times.This will be an important step towards an alternative to potentially harmful synthetic colors. Natural color is one of the crucial factors for the consumer's acceptability of any processed foods. Wine waste material is supposed to be a very potential source of coloring pigment anthocyanin. Natural colors can be extracted by utilizing novel techniques which include soiled-liquid, dynamic superheated liquid extraction, ultrasonics and pulsed electric fields techniques. Anthocyanin is anatural additive for food, cosmetics and pharmaceutical. It is also a good source of antioxidants. The pigments could be shiny orange, pink, red, violet and of blue colors in many flowers and fruits which come under the common class of phenolic compounds (Sulistyo *et al.*, 2016).

Recently, in the food industries much emphasis is given to the production of value added products from the waste generated by agricultural industries. The wastes produced from the agriculture could be anything starting from seeds, peels, stalks, stems or leaves. These contain many biologically active compounds which mostly contain very important antioxidant polyphenols. It has application in the pharmaceutical, cosmetic and food industry. Grape (*Vitisvinifera*) is one of the largest fruit crops in the whole world. Wine, grapes and seed extracts are major sources of antioxidants. They reduce the risk of chronic health disorders including cancer and coronary heart diseases. During wine making from grapes a lot of pomace and seeds are wasted. This can be utilized as a functional ingredient in different food items. This could also provide health benefits. These anthocyanins play a major role as natural colorants. The interest for natural colors is increasing in recent times. It is mainly because of the wide range of colours and beneficial health effects. Anthocyanins have a number of applications in food, pharmaceuticals and cosmetic industries. Anthocyanins have an added advantage as they impart natural color. The food colorant industries have a lot of interest towards this as colors are added to many food products. From time immemorial anthocyanins were used in traditional herbal medicines and also in food. They are also useful against many diseases like hypertension, pyrexia, liver disorders, dysentery, diarrhea, urinary problems and the common cold. Therefore, these are used as a functional food ingredient into our diet. Leticia *et al.*, (2009) studied the antioxidant activity, phenolic compounds and anthocyanins content of eighteen strains of Mexican maize. As anthocyanins are allowed by FDA in foods the consumer demand has increased. Currently, naturally colored blue tortillas are much in demand where anthocyanins are being used. These can be extracted from plants by different means namely, solid-liquid, dynamic superheated liquid extraction, ultarsonics and pulsed electric field techniques.

The procedure of these techniques is as under

1) Solid-liquid extraction techniques

This extraction method is very common for the extraction of many compounds especially from fruits. For the phenolic compounds the extraction is done by grinding, drying or lyophilizing the fruits. It is also done by soaking the fresh fruits and then extracting in solvent using aqueous mixtures of ethanol, methanol or acetone as anthocyanins which are polar in nature. Among the most common methods are those which use acidified methanol or ethanol as extractants. It has been found that the extraction with methanol is very efficient. When anthocyanins were extracted from grape pulp, the extraction with methanol was 20% more effective than with ethanol and 73% more effective when only water was used. Nevertheless, in food industry ethanol is preferred due to the methanol toxicity.

2. Dynamics super heated liquid extraction techniques

Super heated liquid extraction (SLE) has been successfully applied to the extraction of different phenolic compounds like catechins and proanthocyanidins from grape seeds and trans-resveratrol from grapes. In this procedure the temperature is raised above the boiling point of the solvent which increases the diffusion rate, solubility and mass transfer of the compounds and thus decreasing the viscosity and surface. In this method the extent of extraction is very high due to the contact of the compounds with the solvent. The amount of solvent consumption is also less as compared with conventional industrial methods. As light and air are absent there is significant reduction in the degradation and oxidation of these compounds during extraction (Miraje *et al.*, 2015).

3. Ultra sonic assisted extraction techniques

When ultrasound is used with different frequencies ranging from 35 KHz the permeability of cell wallis increased which produces cavitations. When the cavitation breaks down of the cell membrane the internal matter (color and oil) comes out. This extraction is carried out in an ultrasonic bath (Sonorex Bandelin RK 100H) with heating frequency of 35 KHz (Schalltec, Morfelden-Walldorf, Germany) at 70°C held during 1 hr. According to a study the extraction from winery waste material was carried out in a water bath incubated at a temperature of 70 °C held during 3 h. In this case, the solid/liquid ratio was kept at 1:20.

4. Pulsed electric field extraction technique

In this method a current is applied using a Pure Pulse. An exponential decay pulse generator is used which has a voltage of 10 kV and a maximum power of 8 kW. According to the literature with a pulse voltage of 9 kV, an electric field strength of 3 kV cm"1 is obtained. When a series of 30 pulses were applied at ambient temperature a specific energy input of 10 kJ kg"1 is obtained. After the treatment the temperature increase was found to be less than 3°C. In a study a parallel plate treatment chamber consisting of stainless steel electrodes with an electrode area of 140 cm 2 and a gap of 3 cm was used. After every 2 Hz the pulse was repeated. The total treatment time was 15s, for filling and unfilling of the sample. The time required was 1 min. The subsequent extractions were performed at 70 °C which was held for 1 h in a shaken Erlenmeyer flask. Vicas *et al.*, (2016) studied the extraction efficiency of polyphenols with antioxidant capacity from two varieties of grapes (white-Muscat Ottonel (MO) and red-Pinot Noir (PN)) using pulsed electric field (PEF). Grapes were harvested in the Crisana Santimreu vineyard in 2015. The grapes were declustered and crushed and were divided into three parts. A part was treated in PEF, the second one was homogenized, and then was treated in PEF and the last part remained untreated, representing the control samples. The bioactive compounds, like total polyphenol content and total flavonoid content, were determined by the spectrophotometric method. The antioxidant capacity of the samples was determined by three methods (DPPH, FRAP and TEAC). The results showed that the application of PEF to the crushed grapes increases the total polyphenol content 2.28 and 3.15 times and the total flavonoid content 7.17 and 5.29 times for MO and PN, respectively. Also, the antioxidant capacity was significantly increased when it was compared with the control sample. When the treatment in PEF was applied on homogenized grapes both bioactive compounds and antioxidant capacity were significantly increased compared to the samples treated in PEF for crushed grapes. The application of PEF to homogenized grapes was found to be more efficient.

B. Anthocyanins and Health Benefits

Araceli Castaneda-Ovando *et al.*, (2009) studied the chemistry of anthocyanins. It was Maria *et al.* (2012) who studied the antioxidant activity and consumer acceptance of grape seed flour-containing food products. Scientists Musee *et al.*, (2005) studied how the yields of wastes can be minimized during wine-making processes. Anthocyanins and their variation in different red wines has been well studied by Fei He *et al.*, (2012). Ghosh and Konishi (2007) studied the role of anthocyanins and anthocyanin-rich extracts in diabetes and eye function. Izabela and Zhang (2004) studied the formation of natural colors from anthocyanins. Corrales *et al.*, (2008) extracted anthocyanins from grape by-products using ultrasonics, high hydrostatic pressure or pulsed electric fields. Luque-Rodrýguez *et al.*, (2007) extracted anthocyanins and other phenolics from red grape skins of wine making residues using dynamic superheated liquid extraction system. Shipp and Abdel-Aal (2010) studied the food applications and the physiological effects of anthocyanins as functional food ingredients.

4. e. Alkaloids

Residues from growing and processing raw agricultural products such as grain, fruits, poultry, fish and trees can be utilized in agriculture and for agro-industries (Detroy and Hesseltine, 1977).

In agriculture tackling of agricultural and agro industrial waste management has become very important. The various ways in which the agricultural residues can be utilized and which technology is worth taking for the production of single cell protein, methane, organic chemicals and other bioproducts. Technologically and from economic consideration the most important residues are animal wastes, cellulose material, cereal starch and crop residues. Agricultural residue bioconversion is picking up in recent times. Opiate alkaloids can be biotrans formed by microbes and plants. This was well reviewed by Rathbone *et al.*, (2001). The review gives a good summary of the progress of alkaloid biotransformations from mid-1980's to 2002. It is difficult to modify structures of alkaloids because of the complex polycyclic nature. Thus, biotransformation has played a role in generating new and more active derivatives. Microbial hydrolysis can deliver very high yields. Hegazy *et al.*, (2015) found that cinobufagin and resibufogenin could be completely metabolized by *Alternaria alternata* AS 3.4578 which generates 12b-hydroxylated products in more than 90% yield in duration of 8 h. *A. alternata* could also convert 3-epi-desacetylcinobufagin into 3-epi-12b-hydroxyldesacetylcinobufagin as the major product (70%). In addition, four dehydrogenated products,3-keto-resibufogenin, 3-keto-cinobufagin,3-keto-deacetylcinobufagin and deacety-lcinobufagin, were obtained from the biotransformation of resibufogenin and cinobufagin by *Pseudomonas aeruginosa* AS 1.860 (Zhan *et al.*, 2003). While with *Mucor polymorphosporus* AS 3.3443 biotransformation of resibufogenin produced 22 different metabolic products. However the yields were found to be low (Ye *et al.*, 2005). Biotransformation of three compounds resibufogenin, cinobufagin and bufalin were studied in *Nocardia* sp. NRRL 5646 (Zhang *et al.*, 2007). The biotransformed products had higher cytotoxic activity than the substrate. There has been biotransformation reactions which showed great regio-selectivity on bufadienolides. Cinobufagin and bufalin were converted to 3-acetyl cinobufagin and 3-acetyl bufalin respectively. With *Cunninghamella elegans* the biotransformation of cinobufagin produced 12a-hydroxybufagin; 11a-hydroxybufagin, 12b-hydroxydesacetylcinobufagin, 3-oxo-12a-hydroxy-bufagin and 12b-hydroxybufagin were produced (Qiao *et al.*, 2007).

Significant amounts of wastes are generated every year from agriculture residues. These need to be utilized properly. Potential supply of US agricultural byproducts from domestic crops might exceed 500 million tons (dry basis). The feasibility of residue utilisation for energy or chemical production depends on a myriad of residue characteristics: availability, supply and current usage. Residues need to be seen considered as resources and not as wastes. For the large quantities of wastes generated it needs be seen that the utilisation should be successful. For this the costs of existing raw materials, maintenance of a quality environment, raw material availability and the usefulness of the end product needs to be studied.

Residue Identification

Recently, in "Food and Agriculture Organisation of the United Nations Symposium and Management of Agricultural and Agro-Industrial Wastes" discussed various issues of wastes like their characteristics, generation and utilisation technology. Scientists have studied the utilisation of available agro-residues by bio converting them to single cell protein. It has been observed that in majority of the crops the residues produced are left in the fields. It is only with sugarcane, vegetables, some fruits and peanuts that there are significant accumulations at specific processing sites.

By two biological processes the wastes are converted into useful products. These are namely biogas by ananaerobic fermentation and biochemical hydrolysis. The biochemical processes produce primarily protein, sugar and alcohol. The anaerobic fermentation works in the absence of oxygen and produces methane gas. Oils and oilseeds produce residues in large quantities.

These include the harvesting and processing of groundnuts, cotton seed, palm oil and olive oil. It is possible to produce activated carbon from the shells of numerous oilseeds. This serves as a very good resource. The utilization of these residues can be done in the lignin recovery, pentosans and cellulosic materials. Palm oil can be a good source of energy. Scientists have worked on the waste material available from processing sugarcane and coffee as a potential source of single cell protein. Recently, newer techniques like e-burning are used for the hulls consisting of cellulose and related polysaccharides. The dried hulls which are known to contain alkaloids are used as mulch in the coffee fields. Hegazy *et al.*, (2015) took microbial biotransformation as a tool for drug development based on natural products from mevalonic acid pathway.

4. f. Terpenoids

Monoterpenes

Hirokazu *et al.*, (1998) have studied the microbial biotransformations by *Glomerella cingulata* of two saturated acyclic monoterpenoids namely tetrahydrogeraniol and tetrahydrolavandulol. Both compounds were hydroxylated regioselectively at the isopropyl group. Tetrahydrogeraniol was transformed to hydroxyl citronellol, while tetrahydrolavandulol was transformed to 5-hydroxytetrahydrolavandulol. Plant pathogenic fungus *Absidia glauca* metabolized the cyclic monoterpene ketone-carvone. After 4 days of incubation, the diol 10-hydroxy-(+)-neodihydrocarveol was formed via (+)-trans-dihydrocarvone and (+)-neodihydrocarveol in 4 days (Demircia *et al.*, 2004). Tanzi *et al.*, (2012) studied that terpenes can be used as green solvents for extraction of oil from microalgae. Agrawal and Joseph (2000 a, b) studied the bioconversion of available alpha pinene to high valued verbenone by *Aspergillus* sp. and *Penicillium* sp. cultures. Agrawal *et al.*, (1999) improved the biotransformation of alpha pinene to verbenol using *Aspergillus* sp. and *Penicillium* sp. by inducing mutations using chemicals and UV radiations.

Sesquiterpenes

Sesquiterpene phenols (S)-(+)-curcuphenol and (S)-(+)-curcudiol were isolated from the Jamaican sponge *Didiscus oxeata* which were converted into useful products by microbial and chemical transformation. Sesquiterpenoid with *Kluyveromyces marxianus* var. lactis (ATCC 2628) on fermentation gave six new metabolites: (S)-(+)-15-hydroxy curcuphenol, (S)-(+)-12-hydroxy curcuphenol,(S)-(+)-12,15-dihydroxycurcuphenol,(S)-(+)-15-hydroxycurcuphenol-12-al, (S)-(+)-12-carboxy-10,11-dihydrocurcuphenol and(S)-(+)-12-hydroxy-10,11-dihydrocurcuphenol.

Incubation for fourteen-days with *Aspergillus alliaceus* (NRRL315) biotransformed to (S)-(+)-10b-hydroxycurcudiol, (S)-(+)-curcudiol-10-one and (S)-(+)-4-(1-(2-hydroxy-4-methyl) phenyl)) pentanoic acid. Using *Rhizopus arrhizus* (ATCC 11145) and *Rhodotorula glutinus* (ATCC 15125) many useful products were formed on biotransformation. *R.oryzae* converted into many useful products. Both fungi transformed into 16-(1-methyl-1-propenyl) erementhol anolide, providing experimental evidence for the biosynthesis of the eremantholide hemiketal unit. Biotransformation studies were conducted on (+)-(S)-ar-turmerone and (+)-(S)-dihydro-ar-turmerone by utilizing the fungus *Aspergillus niger*. Renu Agrawal *et al*., (1996) studied the biotransformation of alpha pinene to flavour volatiles by *A. niger*.

Diterpenes

Microbial transformation of fungus *Debaryomyces hansenii* produced many useful compounds.

Triterpenes

Biotransformation of nigranoic acid along with *Trichoderma* sp. JY-124 produced (Z)-triene-3, 26-dioic acid a useful compound. Microbial transformation of 4-olean-type pentacyclic triterpenes (OPTs), 3-oxo oleanolic acid, oleanolic acid and esculentoside A were studied. After the screening of 12 strains of microbes, preparative biotransformation by two strains of *Streptomyces griseus* ATCC 13273 and *Aspergillus ochraceus* CICC 40330 resulted in the isolation of 10 metabolites (Shah *et al*., 2014).

Tanzi *et al*., (2012) described a green and original alternative procedure for the extraction of oil from microalgae. Terpenes were used from renewable feedstocks for the extraction instead of dangerous solvents like n-hexane. This was done using soxhlet extraction and then removing the solvent from the medium. This method is eco-friendly. Agricultural waste derived from citrus peels is a terpene (d-Limonene) which is of low cost, low toxicity and is biodegradable. It is an economical renewable feedstock. It can be used in a wide variety of applications due to its degreaser quality. Recently, scientists have studied oil extraction from oil-containing materials. The yield and quality of crude oil obtained from the d-limonene extraction were almost similar to those obtained using n-hexane. Alpha pinene is a natural terpene hydrocarbon obtained from gum turpentine, a kind of essential oil distilled from pine gum (Liu *et al*., 2005).

Another renewable resource is gum turpentine which could be used as a solvent. It is utilized in thinning of the oil based paints and also in the production of varnishes. Gum turpentine is usually obtained from pine,a forest tree which produces high quality of α-pinene, β-pinene, pine oil, terpineol and many other terpene compounds. Another important aromatic hydrocarbon is P-cymene which is produced from tree leaf oils (Hattori, 1981). It is a very important intermediate in the chemical industry. It is also used as a solvent for dyes and varnishes, as a heat transfer medium, in fragrances and perfumes. It is also used to mask odor of a number of industrial products. Use of terpenes as a solvent in extraction of oil-containing materials like microalgae will be very important.

4. g. Steroids

Microbial transformation of diosgenin (3b-hydroxy-5-spirostene) using white-rot fungus *Coriolus versicolor* afforded four previously unreported polyhydroxylated steroids, (Wu *et al.*, 2011).

The microbial biotransformation of steroid compounds has been well reviewed by Fernandes (2006). A lot of progress has been made in this direction. Many steroids like bufadienolides are not only natural but they also possess high amounts of anticancer activities.

Hansen *et al.* (2011) studied the analytical methods to determine ten steroid hormones in animal waste manure and agricultural soil using inverse and integrated clean-up pressurized liquid extraction and gas chromatography-tandem mass spectrometry. According to the workers the livsetsock wastes are natural agricultural waste. These have high amount of EDC's to the environment. Compounds like estriol and male androgen testosterone and its derivative and rostenedione are excreted by animals of all species. These hormones are present in faeces and/or urine and reach the environment through the release of animal wastes. Sarmah *et al.*, (2006) in their investigation found out that it could also reach the environment through water, animal waste application onto land, or by direct excretal input while grazing. All farm animals excrete these hormones, however the type and proportion of estrogens may vary from species to species. In an investigation by Erb *et al.*, (1977) in cattle (Bos Taurus) > 90% of estrogens are excreted as 17a-estradiol, 17P-estradiol, and estriol as free and conjugated metabolites, while pigs (Sus scrofa) and poultry (Gallus domesticus) excrete estradiol, estriol and estrone and their respective conjugates. Hanselman *et al.*, (2003) studied the amounts of hormones excreted by various animals and found that it depends on their reproductive stage and the route of excretion which may be either by feaces or urine.

Many animals including cattles excrete steroid hormones as free and conjugated compounds. The activity can be neutralised and they can be removed from the bloodstream via the kidneys if these steroid hormones can be conjugated either to glucuronides (addition of glucose) and/or sulphate (by addition of sulphate) forms. However, much work needs to be done towards free and conjugated metabolites and also how this may be affected by their oestrous cycle, pregnancy and lactation. Many organic compounds are known to get de-conjugated by microorganisms very fast. Looking into this there is a great probability that even for steroid hormones this might happen. Typical levels of steroid hormones in waste from dairy cows are reported as 39 ^g/kg of estrone and about 18 ^g/kg of 17P-estradiol. Typical slurry pit concentrations of 1.5 ^g/L were obtained for 17P-estradiol, while the concentration of estrone was 3-fold higher (Raman *et al.*, 2004).

Ermawati *et al.*, (2007) studied the degradation and behavior of natural steroid hormones in cow manure waste during biological treatments and ozone oxidation. In a study, cow manure waste can be utilized for the degradation of natural steroid hormones using a draw-and-fill process for thermophilic anaerobic digestion. Aerobic treatment of the effluent is done from the anaerobic treatment after diluting it many times with tap water. Hormone degradation was checked by using a yeast two-hybrid assay. Significant reductions were found in the effluent from the biological treatments which could remove the major part. However for total removal ozone oxidation was necessary. Bignell *et al.*, (2016) studied the secondary metabolite arsenal of an opportunistic pathogenic fungus. El-Fouly *et al.*, (2015) studied the biosynthesis of pyocyanin pigment by *Pseudomonas aeruginosa*. Eriksson *et al.*, (1990) studied the degradation of wood and wood components by microbial and enzymatic methods. Bindman *et al.*, (2015) studied how the facile leader peptides could be removed from lanthipeptides by incorporation of a hydroxy acid.

4.h. Natural Phenols, Essential Oils and Auto-Inducers

There are different types of phytochemicals produced from agriculture wastes, but most important are the polyphenols. These are secondary metabolites that have a number of various compounds. These vary in structure and bioactivities like antioxidant, anti-inflammatory and anti-cancer properties (Mona *et al.*, 2015). For maximum recovery of such compounds much R and D is in progress. Scientists are also looking into the use of these compounds in food, pharmaceuticals and cosmetics without producing any waste (Rashad *et al.*, 2011). Scientists are working on the discovery and design of extraction processes to use alternate solvents and renewable natural anti cancer products which are safe and have high quality (Mahmoud and Ali, 2014). This will help in novel extraction techniques. Recently, a lot of emphasis is on the recovery of polyphenolic compounds from plant food industry wastes. Newer and more efficient techniques are solid-liquid extraction. Microwave heating, ultrasounds and high pressure treatments are utilized with a solvent. These newer methods have made the earlier methods obsolete (Patricia *et al.*, 2010, Vilkhu *et al.* 2008). Most of the methods could not be successful at the industrial level due to the toxicity of solvents. Under such cases again the solvents are needed to be recycled (acetone, methanol), state laws (ethanol), increased cost and doubtful efficiency (pressurized liquids, supercritical fluids) (Salimi *et al.*, 2016). Glycerol has been rarely used. Production of secondary metabolites can be enhanced either by the use of precursors or by a biosynthetic enzyme. Sometimes the use of both precursor and enzyme together can further highten the reaction. The inducer molecules are mostly amino acids. Commonly, these are auto-inducers γ-butyrolactones (butanolides) of the actinomycetes, N-acylhomoserine lactones of Gram-negative bacteria, oligopeptides of Gram-positive bacteria and B-factor (3'-(1-butylphosphoryl) adenosine) of *Amycolatopsis mediterranei*. The induction is done through the receptor proteins which takes part in reducing the chemical and morphological differentiation (secondary metabolism and differentiation into aerial mycelia and spores respectively). When a complex is formed with the butanolide then it becomes non- functional. Homoserine lactones of Gram-negative bacteria function at high cell density and are structurally related to the butanolides. These have the ability to increase plant and animal virulence, light emission, plasmid transfer and production of pigments, cyanide and β-lactam antibiotics. In the case of Gram-positive bacteria quorum sensing operates using the peptide pheromones. During this mechanism the secretion happens using an ATP binding casette (ABC transporter), the secreted pheromone is recognized through a two-component signal transduction system via a sensor. It induces the synthesis of proteins that are involved in protein/peptide antibiotic (including bacteriocins and lantibiotics) production, virulence and genetic competence. The B-factor of *A. mediterranei* is an inducer of ansamycin (rifamycin) formation. Feedback regulation is also involved in secondary metabolite control. Antonious *et al.*, (2009) studied the relation between pesticides, food contaminants and agricultural wastes.

Vidya Laxme *et al.*, (2014) studied the synergistic effects of probiotic *Leuconostoc mesenteroides* and *Bacillus subtilis* in malted ragi (*Eleucine corocana*) food for antagonistic activity against *V. cholerae* and other beneficial properties. Basappa *et al.*, (1997) studied the nutritonal composition of fermented ragi (chang) by Phab and defined starter cultures as compared to unfermented ragi (*Eleucine caracanis* G).Studies were done by Surya Chandra Rao *et al.*, (2007) to improve the stability and viability of probiotic lactic acid bacteria *Pediococcus pentosaceus* by supplementing cysteine hydrochloride

(Cys HCl) and ascorbic acid (AA) as oxygen scavengers at different concentrations. The adjuvants were supplemented to the culture in MRS broth stored at 28°C and 18°C for 12 days keeping in mind the milk products like biocurd (patent no: NF482005) prepared with the same culture till the laboratory to be used for a period of 10 days. It was found that Cys HCI and AA were the best for the stability and viability of probiotic lactic acid bacteria. During the study, the decrease in the pH and increase in the acidity in control along with lower growth were attributed to autolysis as visualized by scanning electron microscopy. Releases of proteins were also observed. The increase in growth in the supplemented samples can be correlated in improving the stability and viability by four logs as compared to the control. Shobharani *et al.*, (2006) improved the strain by mutagenesis and optimized the conditions for culture parameters by response surface methodology for lactose tolerance in a novel native culture isolate *Leuconostoc mesenteroides*. Sudha *et al.*, (2006) studied the stability and viability of a local probiotic isolate *Pediococcus pentosaceous* (MTCC 5151) under induced gastrointestinal tract conditions. The cultures in the above said work were isolated in from the waste fermented products.Vanaja *et al.*, (2010) studied the volatile profile of compounds with therapeutic importance produced by probiotic culture strain of *Lb. plantarum* LB cfr.

Volatile Fatty Acids from Anaerobic Digestion of Waste Streams

Lipids are a good substrate for biodiesel production. It is preferred as it is sustainable, non-toxic and energy efficient. Economically oleaginous microorganisms are usually preferred as these are non-edible sources of lipids. These microbes are able to assimilate the carbon source from a medium and have the potential to convert the carbon into lipid storage materials. This can be transesterified to form biodiesel.

Antioxidant activity and fatty acid profile of fermented milk prepared by using a microorganism of fermented food *Pediococcus pentosaceus* was studied by Gayathri and Renu Agrawal (2014). Shobharani and Renu Agrawal (2007) studied the therapeutic importance of volatile compounds produced by *Leuconostoc paramesenteroides*. Shobharani and Renu Agrawal (2008) studied the effect of cellular membrane fatty acids in the stressed cells of *Leuconostoc mesenteroides* sub sp. *Dextranicum* which was a native probiotic lactic acid bacterium. Alibardi and Cossu (2015) studied the composition variability of the organic fraction of municipal solid waste and effects on hydrogen and methane production potentials. Bhatia and Yang (2017) studied the microbial production of volatile fatty acids: current status and future perspectives. Bolado-Rodríguez *et al.*, (2016) studied the effects of thermal, acid, alkaline and alkaline-peroxide pre-treatments on the biochemical methane potential and kinetics of the anaerobic digestion of wheat straw and sugarcane bagasse. Shobharani and Renu Agrawal (2010b) studied the enhancement of cell stability and viability of probiotic *Leuconostoc mesenteroides* MTCC 5209 on freeze drying to be used in food formulation. Gayathri Krishnan and Renu Agrawal (2012) studied the antioxidant activity and fatty acid profile of fermented milk prepared by *Pediococcus pentosaceus*. Bolzonella *et al*., (2007) studied the two-phase anaerobic digestion of waste activated sludge by an extreme thermophilic prefermentation. Short chain volatile fatty acids can be produced through the acidogenic fermentation of organic waste of different origin. These compounds are extremely useful as starting compounds for the chemical industry of biopolymers, such as PHAs, reduced chemicals and derivatives. A good review in the literature has been written on the work done in twenty years of research on anaerobic fermentation of the organic fraction of municipal solid wastes, fruit/vegetable wastes, waste activated sludge and agrowaste, carried out by the same research group, so to generalize the obtained results and compare those with literature data obtained in similar conditions. The yields in terms of volatile fatty acids production per kg of organic matter (as chemical oxygen demand) were evaluated and compared, considering process parameters such as hydraulic retention time, organic loading rate, and temperature. Influence of temperature and hydraulic retention on the production of volatile fatty acids during anaerobic fermentation of cow manure and maize silage (Cavinato *et al*., 2017). Shobharani and Renu Agrawal (2010a) studied the interception of quorum sensing signal molecule by furanone to enhance shelf life of fermented milk.

Antimicrobial Potential of Agro-Industrial Wastes

Some agro-industrial wastes have shown antimicrobial (Doughari *et al*.2008b) and antioxidant activities (Zargham and Zargham 2008). Wastes from green tea processing, acorn, chestnut and persimmon hulls have been studied for the extraction of tannins, antibacterial and antioxidant activities by Sung *et al*., (2012). It has been observed that tannin content varies with the solvent used. Therefore, it was analysed using three solvents namely ethanol, acetone and aqueous extracts. It was observed that tannins were present in the extracts of chestnut hull, green tea waste, acorn hull and persimmon. Statistics show that large amount of agricultural wastes have tannins in the form of polyphenols. In the solvents tested maximum tannin was found in the ethanol extract of green tea. In chestnut hull it was highest in ethanol and acetone extracts. Antibacterial activities of various extracts were screened against *Staphylococcus aureus, E. coli, S. flexneri, L. monocytogenes* and *B. coagulans*. Even the antibacterial activity was highest in the tannin extracts of green tea waste. Renu Agrawal and Shylaja Dharmesh (2011) studied an antishigella protein from *Pediococcus pentosaceous* MTCC 5151, a local isolate from fermented food. Vanitha and Renu Agrawal (2012) purified a protein from probiotic *Leuconostoc mesenteroides* which was found to be active against *V. cholerae*.

Commonly known, Jojoba *(Simmondsia chinensis)* is commercially cultivated in many countries all over the world. It is mainly a commercial crop used for its seed oil. Some animals also use it as food. The oil has many useful properties which are utilized in medicines. It is widely used in the treatment of cancer, kidney disorder, obesity, sore heart, warts and wounds (Leung, 1977). Earlier Wagdy and Taha (2012) have studied the antimicrobial activity of phenolic extracts of jojoba hulls solvent extraction.The antimicrobial activity was tested using pathogenic microorganisms as *Escherichia coli,Staphylococcus aureus, Bacillus cereus, Listeria monocytogenes* and *Salmonella typhimurium*. The solvents used for the extraction were methanol, ethanol, acetone, isopropanol and ethyl acetate. The extracts obtained from different solvents of jojoba hulls had varying antimicrobial activity. Ethyl acetate extracts showed the maximum inhibition in *B.cereus* and *S.typhimurium*. Whereas for *S. aureus*the antimicrobial activity was found highest in methanol extract and in *L. monocytogenes* and *E.coli* it was maximum in ethanol and acetone extract respectively. With their investigation, Wagdy and Taha (2012) came to a conclusion that jojoba hull was a good source of bioactive compounds with antioxidant, antimicrobial, and anticancer properties which are very good. Rao and Mikkelsen (1977) were successful on production of organic acids in flooded soil using rice straw.

Work by various scientists (Ragazzi *et al.*, 1973; Angerosa *et al.* 1995; Romero *et al.*, 2002) has shown that olive fruits have very high content of phenolic compounds which include phenol acids, phenol alcohols, catechol, methyl catechol, phenyl alcohols (tyrosol, hydroxytyrosol), flavonoids (luteolin-7-glucoside, apigenin-7-glucoside, rutin and quercetin). It also includes many anthocyanin pigments (cyaniding-3-glucoside and cyaniding-3-rutinoside). Hydroxytyrosol is one of the major phenolic compounds present in olive fruit with remarkable pharmacological and antioxidant activity (Visioli *et al.*, 2006; Fernandez *et al.*, 2006). Works done by scientists (Aziz *et al.*, 1998; Visioli and Galli 2002) have shown that oleuropein is another valuable compound that has antiviral, antibacterial, antifungal, antioxidant and anti-inflammatory properties. Olive phenolic compounds have antibacterial activities against many pathogenic bacteria like *Salmonella* spp., *Staphylococcus aureus*, *Clostridium botulinum* and *Listeria monocytogenes* (Payne *et al.*, 1989; Nychas *et al.*, 1990; Tassou and Nychas 1995). Olive oil mill waste contains polyphenols in large quanities. This inhibits the growth of pathogenic bacteria like *Pseudomonas syringae* and *Corynebacterium michiganense* (Capasso *et al.*, 1995). Even the olive oil mill waste water (OMWW) had poly phenols. This was studied by Ciafardini and Zullo (1998) on the crucifer seed-borne phytopathogen *Xanthomonas campestris*. The mechanism of action was on the protein of the bacterial cell walls which disrupts them. OMWW was found to inhibit the seed-borne phytopathogen *Xanthomonas campestris* completely without damaging the germination ability of the crucifer seeds and the metallic greenhouse structures. Ciafardini and Zullo (2001) studied theâ-glucosidase activity in *Leuconostoc mesenteroides* associated with fermentation of "Coratina" cultivar olives. Many scientists worked on the antimicrobial activity from other fruits. Martin *et al.*, (2012) assessed the antimicrobial potential and chemical composition of guava bagasse (*Psidium guajava*), Pinot Noir (*Vitis vinifera*) grape marc, Isabella grape marc (*Vitis labrusca*) wastes, Petit Verdot Verdejo grape marcs, Syrah and Verdejo grape stems, Petit Verdot grape seed and red grape fermentation lees (*Vitis vinifera*) and tomato bagasse (*Solanum lycopersicum*) wastes. Work was done on vegetable wastes like kale (*Brassicaoleracea*), beet (*Beta vulgaris*), broccoli (*Brassica oleracea*), turnip stems (*Brassica rapa*), carrot (*Daucus carota*), radish leaves (*Raphanus sativus*), pumpkin (*Cucurbita* sp.), peanut peel and passion fruit hulls (*Passiflora edulis*) against pathogenic microorganisms. For extraction in solvents they were immersed in ethanol and methanol solutions. Beet stalk, peanut peel, Pinot Noir grape marc, Petit Verdot grape seed and marc, red grape fermentation lees and guava bagasse wastes showed antimicrobial activities against Gram-positive bacteria *Staphylococcus aureus* and *Listeria monocytogenes*. Methanol extract of peanut peels and ethanol extract of guava bagasse extracts showed the lowest MIC against *S. aureus* and *L. monocytogenes*. Guava bagasse extract showed the highest antimicrobial activity against *L. monocytogenes*.

A food industry today needs to replace the synthetic preservatives and is looking for newer natural compounds which are safe and can prevent contamination. The best alternative is thus the agro-industrial wastes. These are rich in many bioactive compounds. This will also help in green environment. Therefore, Martin *et al.*, (2012) assessed the antimicrobial potential and chemical composition of agro-industrial wastes against pathogenic microorganisms of importance in foods. Beet stalk, peanut peel, Pinot Noir grape marc, Petit Verdot grape seed and marc, red grapes fermentation lees and guava bagasse wastes showed antimicrobial activity against *Staphylococcus aureus* and *Listeria monocytogenes*. The minimum inhibitory concentrations ranged from 0.78 to 25mg/ml. Wastes with antimicrobial activity showed the highest total phenolic compounds among the wastes studied (37.3 to 400.2g GAE/kg). A GC-MS profile was also studied that identified compounds like caffeic, gallic, ferulic and ñ-coumaric acids. Presence of flavonoids like quercetin, myrcetin and epicatechin from the wastes exhibited antimicrobial activity. They also demonstrated that agro-industrial wastes from wine and food processing industries could be used as a source of research on new antimicrobial compounds for use by food and beverage industry as natural preservatives. Mandalari *et al.*, (2006) studied and characterized the flavonoids and pectin from bergamot (*Citrus bergamia Risso*) peel, a major byproduct of essential oil extraction. Mandalari *et al.*, (2007) studied the antimicrobial activity of flavonoids extracted from bergamot (*Citrus bergamia* Risso) peel, a byproduct of the essential oil industry.

4. i. Phenazines

There are many bioplastics which are produced from agricultural waste materials (Xu *et al.*, 2012). However, these bioplastics have very high cost due to the high prices of the feedstocks and agricultural raw materials. To be competitive, the prices of the raw materials needs to come down. Bioplastics can be produced by the fermentation of lignocellulose, molasses, whey, oils and other wastes and thermoplastics from natural macromolecules via chemical modifications. The scientists produced bioplastics from waste materials and low-value byproduct, degradable polymers and materials. *Pseudomonas* strains (Sudhakar *et al.*, 2015) were isolated from the soil, water and clinical specimens. Most of them were *Pseudomonas aeruginosa* which produced pyocyanin. Some were isolated from ricecultivated soil and some from clinical specimen of urinary tract infection. It was found that these produced highest amount of pyocyanin. Pyocyanin production reached 9.3 and 5.9 μg/ml, respectively on synthetic glucose supplemented nutrient medium (GSNB). The scientists identified these strains (*P. aeruginosa* R1 and *P. aeruginosa* U3) by 16S rRNA. It was also observed that the similarity with other available strains in the database was 97% (with *P. aeruginosa* FPVC 14) and 94% (with *P. aeruginosa* 13.A), respectively. *P. aeruginosa* R1 and *P. aeruginosa* U3 are accessed at gene bank with accession numbers KM924432 and KM603511, in the same order. In this work they extracted pyocyanin by standard methods, purified by column chromatography and characterized by various methods like UV-Vis absorption, mass spectrometry and nuclear magnetic resonance. Many multi-drug resistant microbes are creating problem. Activity was investigated in pure pyocyanin. It was found to be more in Gram+ve bacteria than Gram-ve bacteria and yeast. To reduce the cost of pyocyanin production, a new conventional medium based on cotton seed meal supplemented with peptone was designed. The pyocyanin production from both selected strains *P. aeruginosa* R1 and *P. aeruginosa* U3 using the new medium was found to increase by 30.1% and 17.2%, respectively in comparison with synthetic GSNB medium, whereas the cost of production process was reduced by 56.7%.

4. j. Polyketides

Lactones

Biotransformation of Artemisinin, a sesquiterpene lactone

Artemisinin is a sesquiterpene lactone which has an endoperoxide bridge. It was isolated from the Chinese herbal plant, *Artemisia annua* L. in 1972 (Mahomoodally *et al.*, 2013). It has high therapeutic value in treating malaria. Towards this, studies have been done using *A. niger* AS 3.795 (Mohamed *et al.*, 2015). Work has been done on the biotransformation products of artemisin in for anti malaria (Klayman, 1985). Many oxidative products of artemisinin have been also reported. A novel ketone derivative of artemisin in has been biotransformed by *Streptomyces griseus* ATCC 13273 (Liu, 2006). Bark of *Taxus brevifolia* is very useful as it contains paclitaxel which can be biotransformed into taxol and other compounds. Taxol is one of the most effective anticancer agents obtained from a natural source. It has been found to be very useful in the treatment of ovarian, breast and lung cancers. To enhance the oral bioavailability (500 microorganisms) it was screened for the ability of biotransformation of taxol derivative (cephalomannine). Taxol/cephalomannine was biotransformated by *Streptomyces* sp. MA 7065, yielding 10-hydroxyacetyl-10-deacetyltaxol (116), 30-(4-hydroxy phenyl)-30-dephenyltaxol and 400-Hydroxycephalomannine (Shieh-Shung Tom Chen and Ching-Jer Chang, US 5756536; 1998). They worked on a biotransformation process for the production of hydroxylated derivatives of taxol and new derivatives of taxol and cephalomannine are described. A method for converting taxol to L-760,006 and L-760,007, and a method for converting cephalomanine to L-745,596 are provided. Chen Shieh-Shung Tom and Chang (1998) studied the microbial transformation of taxol and cephalomannine. Lee *et al.*, (2015) isolated panaxosides or ginsenosides from the roots of *Panax ginseng*. These are very effective constituents isolated from the traditional Chinese herb ginseng. Metabolites of ginsenosides are responsible for the major pharmacological activities of ginseng roots (Fu *et al.*,1994). Investigation on antitumor activities of 20(S)-protopanaxatriol showed that it did not directly inhibit tumor growth *in vivo*, but that it stimulated splenic NK cells to become cytotoxic to tumor cells (Shin *et al.*, 2004). There area number of reports in literature on the biotransformation of ginsenosides. The ginsenosides 20(S)-Protopanaxatriol was transformed by a fungus *Mucor spinosus* AS 3.3450. Many novel compounds could be produced from it. These include 12-oxo-15a-hydroxyl-20(S)-protopanaxatriol. Biotransformation offers a versatile tool for structural modification of alkaloids in addition to known chemical methods. Microbial metabolism of evodiamine by *Penicillium janthinellum* and its application for metabolite identification in rat urine was studied (Li *et al.*, 2006). The microorganisms and rats used similar metabolic pathway for evodiamine. Many fungi have inherent enzymes which catalyze alkaloids (Abel *et al.* 2003). It has been investigated that *Mucor piriformis* selectively forms N-demethylated thebaine whereas *Streptomyces* and *C. echinulata* strains form N-demethylated indole alkaloid lergotrile (Ye *et al.*, 2005). Microbial strains of *Cunninghamella* and *Fusarium* strains form N-demethylated codeine (Qiao *et al.* 2007). Biotransformation of a thebaine derivative using the filamentous fungus *C. echinulata* NRRL1384 has also been worked out. The thebaine analog was converted to a mixture of N-demethylated and N,O-demethylated products (Barrero *et al.* 1999). Maiti *et al.*, (2018) studied the methods for enhanced biobutanol production from agro-industrial wastes by hydrolytic pre-treatment.

4. k. Fatty Acid Synthase Products

In this process a supercritical procedure is involved. This is a catalyst-free method and transesterification is done using supercritical methanol at high temperature and pressure in a continuous process. In the supercritical state the oil and methanol are in a single phase and reaction occurs spontaneously and rapidly (Metreand Kaushik, 2013). The process can tolerate water in the feedstock. Free fatty acids are converted to methyl esters.There is no saponification. In this method wide variety of feedstocks can be utilized. There is no need to remove the catalyst. However, high temperatures and pressures are needed. It was found that the energy costs of production were similar or even less than incurred in catalytic production routes. A thorough study was done by Garrido *et al.*, 2015) in which a study was done on the fatty acid (FA) metabolism. The scientists found that the fatty acids were deregulated in several human diseases including metabolic syndrome, type 2 diabetes and cancers. This shows that FA-metabolic enzymes are the potential targets for drug therapy. However, it is important to recheck the results. They found that in healthy organism, synthesis of triacylglycerols (TAGs) composed of three FA units esterified to a glycerol backboneis increased in response to dietary sugar. Saturation in the storage and synthesis capacity of TAGs is associated with type 2 diabetes progression. Sugar toxicity likely depends on advanced-glycation-end-products (AGEs) that form through the covalent bonding between amine groups and carbonyl groups of sugar or their derivatives α-oxoaldehydes. Methylglyoxal (MG) is a highly reactive α-oxoaldehyde that is derived from glycolysis through a non-enzymatic reaction. Glyoxalase 1 (Glo1) works to neutralize MG, reducing its deleterious effects. They found that FASN mutants were lethal but can be rescued by an appropriate lipid diet. Rescued animals did not exhibit insulin resistance, and were very sensitive to dietary sugar and they accumulated AGEs. They could prove that FASN and Glo1 cooperate at systemic and cell autonomous levels to protect against sugar toxicity. It was also found that the size of FASN mutant cells decreased as the dietary sucrose increased. Genetic interactions at the cell-autonomous level, where glycolytic enzymes or Glo1 were manipulated in FASN mutant cells, revealed that this sugar-dependent size reduction is a direct consequence of MG-derived-AGE accumulation. Zhang and Taihua (2011) optimized pectin extraction with the residues of sweet potatoes.

4. l. Non ribosomal Peptides

There is a class of peptide secondary metabolites which are produced by microorganisms like bacteria and fungi and are nonribosomal peptides (NRP). Nonribosomal peptides found in higher organisms are made by the bacteria present inside these organisms (Saini *et al.*, 2014). Nonribosomal peptides are synthesized by an enzyme known as nonribosomal peptide synthetases. These are independent of messenger RNA. Each nonribosomal peptide synthetase can synthesize only one type of peptide. These are usually cyclic and/or branched structures that might contain non-proteinogenic amino acids. They can modify N-methyl and N-formyl groups, can be glycosylated, acylated, halogenated, or hydroxylated. Nonribosomal peptides are dimers or trimers having identical sequences which are chained together, cyclized or branched. Nonribosomal peptides are a very diverse family of natural products with different types of biological activities and pharmacological properties. They are often toxins, siderophores or pigments. Nonribosomal peptides are commercially utilized in antibiotics, cytostatics and as immunosuppressants.

Bignell *et al.*, (2016) studied the secondary metabolite arsenal of an opportunistic pathogenic fungus. *Aspergillus fumigatus* is a very versatile fungus. It is found in agricultural waste products. It has many genetic loci. These contain biosynthetic gene clusters (BGCs) which produce bioactive small molecules (secondary metabolites/natural products). These BGCs are known to enhance virulence. These may cause white-nose syndrome in bats. There is an excellent review written by Lemes *et al.*, (2016) where they have described bioactive peptides as the new generation biologically active regulators. These prevent the mechanism of oxidation and microbial degradation in foods. It has also been found that they increase the treatment of many diseases and disorders, thus increasing the quality of life. They have discussed on the formation of the bioactive peptides from residual waste and other high valued products. Newer nanotechnology has been discussed for the encapsulation, protection and release of controlled peptides and large-scale recovery and purification of peptides aiming at future applications to pharmaceutical and food industries.

4. m. Ribosomally synthesized and post translationally modified peptides

Automated genome mining of ribosomal peptide for natural products has been well studied by Mohimani *et al.*, (2014). From microbial sources a large group of bioactive natural products are produced. These are mainly ribosomally synthesized and are modified peptides (RiPPs) after translation reaction.In view of increased microbial genome databases and improved mass spectrometry (MS)-based metabolomic platforms it is found that today there is a need for computational tools that can connect the genotypes of these natural products predicted from microbial genome sequences with their corresponding chemo types from metabolomic data sets. Recently scientists have come out with innovative method for identification of microbial RiPPs which is known as RiPPquest, a tandem mass spectrometry database search tool. This can be applied to lanthipeptide discovery. RiPPquest uses genomics to limit search space to the vicinity of RiPP biosynthetic genes and proteomics to analyze extensive peptide modifications. The p-values can be computed using peptide-spectrum matches (PSMs). The work has mainly focused on the RiPP quest by connecting multiple RiPPs from extracts of *Streptomyces* to their gene clusters and by the discovery of a new class III lanthipeptide, informatipeptin, from *Streptomyces viridochromogenes* DSM 40736 to reflect that it is a natural product that was discovered by mass spectrometry based genome mining using algorithmic tools rather than manual inspection of mass spectrometry data and genetic information.

Liujie and Wilfred (2016) studied the characterization of Bicereucin, an unusual d-amino acid containing mixed two component lantibiotic. A new group of microorganisms are known as lantibiotics which is a group synthesized by the ribosomes and has modified peptides after translation (RiPPs). On screening it is found that they have antimicrobial activity. The specific presence of these microorganisms is by the thioether present in the bisamino acids lanthionine and methyl lanthionine. The study has reported a two-component lantibiotic from *Bacillus cereus* SJ1 which has unusual structural features that were named as bicereucin. Unlike all previous two-component lantibiotics, only one of the two peptides of bicereucin contains a lanthionine. The second peptide lacks any cysteines but contains several d-amino acids. These are installed by the dehydrogenase BsjJB, the activity of which was successfully reconstituted *in vitro*. The proteolytic removal of the leader peptide was also performed *in vitro*. Bicereucin displayed synergistic antimicrobial activities against Gram-positive strains including methicillin-resistant *Staphylococcus aureus* and vancomycin resistant *Enterococci* as well as hemolytic activity. To illustrate the utility of the enzymes, an analog of the d-amino acid containing opioid dermorphin was successfully produced in *E. coli* by employing the dehydratase BsjM and the dehydrogenase NpnJA.

Bindman *et al*., (2015) showed how facile leader peptides could be removed from lanthipeptides with the incorporation of a hydroxy acid. For the biosynthesis of ribosomally synthesized and post-translationally modified peptide (RiPP) natural products, it typically involves a precursor peptide which contains a leader peptide that is important for the modification process. This is removed in the final step by a protease. It has been observed that the genome mining for new RiPPs is usually hampered as there is no way to remove the leader peptides. The above work has shown the incorporation of hydroxy acids into the precursor peptides of *E. coli* which results in connection of the leader peptide via an ester linkage that is readily cleaved by simple hydrolysis. This has been worked out in two organisms' lacticin 481 and nukacin ISK-1.

4. n. BioEnergy

In case of usage of agricultural crop waste a disposal cost is required. The conversion processes especially for heat, power generation and transport of fuel production should be economical and must have market potential. They are widely used in countries like Sweden, Denmark, Netherlands, USA, Canada, Austria and Finland. There are many benefits with the innovations in agricultural practices. This has given increased biomass yield and the cultivation costs have come down. As these are green technologies they definitely improve the environment. The energy potential of biomass feedstock is also increasing with more research being carried out in the fields of plant genetics, analytical techniques, remote sensing and geographic information systems (GIS). In case agriculturists use these various biological and thermal processes (anaerobic digestion, combustion, gasification, pyrolysis) the wastes can be converted into biofuels, heat or electricity. This will help towards sustainable development in rural areas in a multifaced manner uplifting the rural folks economically, socially and environmentally. In many parts of India electricity has still not reached and people undergo a lot of problems. Biomass energy will definitely reduce 'fuel poverty'. This may bring economic development. Carmona *et al*., (2012) studied the composting from the wastes of wine industry that can be utilized as a substrate for growing soilless ornamental plants. Castaño *et al*., (2011) studied that the organic matter fractions have the ability to suppress *Fusarium* wilt in organic growth media.

4. o. Nano materials

Parisi *et al*., (2014, 2015) studied the current trends in agricultural nanotechnologies. European Commission has put nanotechnology as one of the key enabling technology. According to EC it can contribute to sustainable competitiveness and growth in many industrial sectors. A lot of research has been done in this direction but the practical applications are few. Moreover, it needs to be seen if agricultural nanotechnology can give sufficient economic returns as the initial investments are very high. New nanotech regulation in the EU is bringing regulatory uncertainty for nano products. In recent times, public has become aware of the benefits of nanotech products. This will bring in more consumer acceptance for its applications. The utility of nanotechnology will be transferred to agricultural applications apart from other key industries which will improve their development. Claudia *et al.*, (2015) have written an excellent article on agricultural nano technologies. CNST Nanotechnology Workshop (2013) at Illinois discussed on different nano techniques to be adopted in agriculture. Hirokazu *et al.*, (2015) have discussed on the current possibilities of nano technologies in agriculture. Parisi *et al.*, (2015) studied on the current possibilities of agricultural nanotechnology.

Nanotechnologies and nanomaterials like electronics and personal care products are coming into the market. These nanomaterials (which are just several millionths of a millimetre in size) have special chemical and physical properties. These are highly efficient electrical conductors, while others are used to create water-repellent or dirt repellent coatings. Due to this, the use of nanomaterials have entered many sectors like the consumer, industrial and medical sectors. Some examples include sunlight resistant paints, light weight plastics for the automotive and aircraft industries and antibacterial coatings. Nanomaterials are also utilized in cosmetics, preservatives, cleaning agents, food packaging and textiles. Exposure to nanomaterials causes risks to health. It was in 2006 that Health advisory council had reported about it. However, a lot more evidence is required in this direction. There is a need to have methods to measure as to what is the concentration of nanomaterials involved in the system. Although evidences are required but it is better not to expose human beings to nano materials. Among the waste materials it is important to have a check on nanomaterials. Waste management of nanomaterials is important. This can be done by recycling and incineration of solid waste and waste water purification. Nanotechnology is mainly applied to agricultural production.

Recently (Parisi *et al.*, 2014) dealt mainly with the agricultural sector. The research was undertaken towards the development and possible marketing of the products.

Like nanocapsules, nanoparticles and viral capsids are a part of nano technology that helps in the detection of diseases and treating them. The nutrient absorption by the plants can be increased by this method. It also helps in delivering the active ingredients to specific sites. The use of target-specific nanoparticles can reduce the damage to the plant tissues resulting in the reduction of the levels of chemicals which are otherwise released into the environment. These are being exploited in the field of plant breeding and genetic transformation. Nanoparticles which are obtained from biopolymers like proteins and carbohydrates are found to be good and do not have harmful effects on human health or environment. Recently, scientists are working on the potential of starch-based nanoparticles which are nontoxic. They are good for agrochemicals and are biostimulants.

Electrochemically active carbon nanotubes, nanofibers and fullerenes have unique chemical, physical and mechanical properties. They are applied in highly sensitive bio-chemical sensors. These nanosensors can be applied in agriculture for soil analysis, easy bio-chemical sensing and control, water management and delivery, pesticide and nutrient delivery. Useful nanocomposite processes are being obtained from traditionally harvested materials.

In the agricultural sector it has been observed that the technology oriented nano products like soil enhancer products promote water distribution and storage. It has been found that consequently water can be saved. This technique has been launched by some companies. Due to the high cost they have been applied only at small scale till now. It is estimated that it might be cost effective in the medical or pharmaceutical fields. A lot of rapid research is going on in agro-chemical sector to look into the economics. Companies are also facing challenges derived from the definition of nanomaterials that is adopted by the EU. Nanoscale formulants (clay, silica, polymers, pigments, macromolecules) have been in use in daily household products. Many countries like Australia/New Zealand, Canada, China, the EU, Japan, Switzerland and the US are looking into the risk of using nanomaterials in the food and agriculture sectors. Till now except the EU, no country has made any regulatory framework for the mandatory labelling of nanomaterials in food. Nanotechnology finds applications in many different fields and nano-based inventions have been granted patents. Till now literature finds more than 3,000 patents worldwide using nanotechnology which have potential as agrochemicals. In developing countries it has application in the areas of agri-food, food security, input delivery, rice production systems, agri-biotechnology, healthcare of animals, precision farming, food industry and water use. If the technology becomes cost effective it will find much use.

4. p. Compost

Singh and Nain (2014) studied the role of microorganisms in the conversion of agricultural wastes to compost. The role of microorganisms in the recycling of agricultural wastes is inevitable. After the bioconversion of agro residues there is generation of compost that has many benefits. It increases the soil fertility and soil health. This in turn improves the agricultural productivity, soil biodiversity, reduces ecological risks and creates to heal thier environment. This helps in processing of the enormous quantities of agrowastes with minimum environment pollution. Survase (2012) studied the feasibility of bioconversion of agro waste into edible protein rich mushroom by *Pleurotus sajor* caju (FR.). Recently, people have tried biocompost in the use of bioinoculants as biocontrol agents to inhibit diseases and also for bioremediation. High quality compost can be prepared by using appropriate microorganisms. A lot of research is being carried out in this direction. There is also research required for sustainable technologies for large-scale production. It is observed that paddy straw increases the CH_4 emission especially in irrigated soils (Boateng *et al.*, 2017).This gas increases the global warming. The action of lignocellulolytic microorganisms helps in making compost from different types of agricultural wastes. This also helps in cheaper cost. When utilised in soil it improves soil health. During the making of compost the organic substrates are degraded. (Pan *et al.*, 2012). In this process the mesophilic microorganisms multiply which results in high temperature. This is followed by increase in thermophilic microbes. This process helps in killing the pathogenic organisms and therefore the seeds can be sown for a good crop. Several fungi like *Trichoderma harzianum*, *Pleurotus ostreatus*, *Polyporus ostriformis* and *Phanerochaete chrysosporium* are known to play an important role in composting of lignocellulosic materials (Paszczynski and Crawford 1995). A review has been written on the composting of lignocellulosic material to obtain good and mature compost (Akhtar *et al.*, 1998).

It has been found by the scientists that when the poultry droppings and rock phosphate (1%) are mixed with the compost the content of N-enriched phosphor gets higher which improves the soil health. Thermophilic fungal consortium of *A. nidulans*, *Scytalidium thermophilum* and *Humicola* sp. were found to be very effective.

Scientists at the Directorate of Mushroom Research, Solan have tried using a consortium of thermophilic microorganisms (*Scytalidium thermophilum*, *Humicola insolens* and *Sporotrichum thermophilum*) to produce compost. It was found that pathogen free compost could be obtained in 10 days which was good to cultivate mushroom. The application of mineral enriched compost with high N and P in nutrient deficient soils helps to grow plants. Bioremediation by co-composting is an economically attractive method for cleaning organic pollutants like xenobiotic, petroleum products, polycyclic aromatic hydrocarbons (PAHs) etc. Microorganisms consume contaminants from the soils, ground, surface waters and air. Bulking agents like peat moss, pine wood shavings, bran flakes accelerate composting by favouring the growth of aerobic microorganisms (Sayara *et al.* 2011; Chidthaisong *et al.*, 1996). In recent works scientists have used inclusion of faecal residues/animal litter, organic/inorganic compounds and consortia of effective microorganisms which results in improved compost. Usage of this compost has resulted in better crop-production, its protection and natural resource management. There are different agro-climatic zones around the world. This requires different types of soil. For this, it is important to have special bioino-culants and better technologies. These may help the farmers. Much work is needed in microbiology during various stages of composting. Advanced techniques like denaturing gradient gel electrophoresis (DGGE) and other cultivation independent techniques are utilized for the unculturable microorganisms and their effect in the composting processes.

4. q. Microbial Bioconversions

From time immemorial scientists have been working on microbial transformations. These reactions help in utilizing low cost substrates into high valued products. As the doubling time of microorganisms is very fast and they also produce many enzymes the reaction yields are high. There are many compounds after biotransformations which can be utilized by the industries and come under the category of natural products. Recently many chemical and pharmaceutical industries are using these compounds. Biocatalysts help in the production process of many compounds used as drug intermediates. Stereo specificity is very important and plays a major role in obtaining any specific compound. In traditional processes the steps involved in separation and purification are very complicated, labour intensive and expensive. All these things can be easily solved by microbial transformations. Biocatalysts work under mild conditions and produce very useful products than by chemical catalyst. Lot more research is underway to find out newer biocatalysts and to maximize the yields. The industries are looking ahead for more effective natural products which are more efficient and are able to produce products with high yields (Hegazy *et al.*, 2015).

Using biotransformation a definite and specific enantiomeric compound of a distinct structure and regio-specificity can be obtained. These processes mostly include microorganisms like bacteria, fungi and yeasts. The bioconversion reactions mostly involve enzymes. The advantage of using microorganisms is that when live microorganisms are used in biotransformations they continuously produce enzymes and the product can be released continuously with homogeneity. They can be produced batch after batch in large amounts. Many chemical and pharmaceutical industries are using them for drug manufacture. (Frydman *et al.* 2005). Cherry and Findatsef (2003) have written a good review on the evolution of industrial enzymes.

Biotransformation processes are very diverse as compared to therapeutic protein production processes (Collins and Kennedy, 1999). For any selective biotransformation the strain of microorganism and the enzymes produced by it has a major role towards maximizing the product production. Biotransformation processes have the advantage as they fall under green chemistry. All governments are looking towards technologies which have minimal waste production and energy consumption (Woodley and Lilly, 1994). Therefore, biotransformation processes play a major role in green chemistry towards the production of compounds especially for pharmaceuticals and drug industry in recent times. As there are many versatile microorganisms these can be utilized for different specific conversions using substrates of low cost (Zheng *et al.* 2014). There are many studies which can be performed to a larger extent regarding different endophyte species for the chemical alterations of molecules and compounds by microbial biotransformations. Divyashree *et al.*, (2006) studied the biotransformation of chiral terpenes by *A. niger* and *Pseudomonas putida*. Vidya and Agrawal (2003) studied the production of verbenol, a high valued food flavourant from improved culture strain of *A. niger* using protoplast fusion. Renu Agrawal *et al.*, (2003) studied the biotransformation of ferulic acid to vanillin by locally isolated bacterial cultures. Smitha Rao *et al.*, (2003) worked for the enhancement of verbenol, a high valued food flavourant by an intergeneric fusant strain of *Aspergillus niger* and *Penicillium digitatum*. Nazhat and Renu Agrawal (2003) studied the solubilization and stabilization of alpha pinene hydroxylase from *Aspergillus niger*. Renu Agrawal *et al.*, (2000) studied the flavour profile of idli batter prepared for defined microbial starter cultures. Renu Agrawal and Joseph (2000a) studied the factors affecting bioconversion of alpha pinene to verbenone by resting cells of *Aspergillus niger*. Renu Agrawal and Joseph (2000 b) optimized the conditions for the biotransformation of alpha pinene to verbenone by a local soil isolate *Penicillium* sp. The biotransformation of alpha pinene to verbenone by resting cells of *Penicillium* sp. was influenced by the conditions of growth of the fungal culture in cultivation medium and the conditions of buffer in which the biotransformation was performed, potato dextrose broth with supplementation of 1% glucose and 0.025% yeast extract of pH 5.75 and a temperature of 30^0C were the conditions for culture growth. Biotransformation was done in buffer. Fungal biomass (200 mg dwt) with 20 mg of alpha pinene in 100 ml of 0.05 M phosphate (pH 7.0) at 30^0C and 6 h of incubation were found to be optimal for the biotransformation. Renu Agrawal *et al.*, (2000d) studied the conversion of nerol to dihydrocarvyl acetate by *Mucor* sp. Renu Agrawal *et al.*, (1999) improved the biotransformation of alpha pinene to verbenol in *Aspergillus* sp. and *Penicillium* sp. which were induced by mutation using chemicals and ultra violet irradiation.

Venkateshwaran *et al.*, (1999) studied the the utilization of catalase using *Saccharomyces cerevisiae.* A culture medium was designed for maximal production of biomass and intracellular catalase by a potent enriched culture of *Saccharomyces cerevisiae* CFR-201. Growth conditions for catalase production were optimal at 30 degrees C for 72 h, in shake flasks on a rotary shaker (200 rpm). The effects of various carbon and nitrogen sources on catalase production were also studied. *S.cerevisiae* CFR-201 produced five times more intracellular catalase than a local baker's yeast strain. The cell free extract used as source of catalase removed 99.9% residual H_2O_2 from cold pasteurized milk at 45 ^{0}C for 1 h.

Basappa *et al.* (1997) studied the nutritional composition of fermented ragi (chhang) by phab and defined starter cultures as compared to unfermented ragi (*Eleucine coracana* G.). Cooked ragi (*Eleucine coracana* G.) fermented separately either by phab or selected starter cultures isolated from phab and their water extracts (chhang) were analysed for various nutritional factors and compared along with that of unfermented but cooked ragi was not subjected to fermentation. No significant nutritional differences were observed between the two types of fermented ragi (chhang). However, in comparison with unfermented cooked ragi, the fermented ones showed increase in protein content (7.6 to 10.5%) and some free amino acids in addition to the synthesis of extra amino acids. More significantly there were higher levels of riboflavin (0.62mg/100g), pantothenic acid (1.6mg/100g) and niacin (4.2mg/100g) in the fermented grain and uniquely cyanocobalamin was synthesised anew (40µg/100g) during ragi fermentation. Significant amounts of bioavailable minerals (Ca, P and Fe) in the beverage were also found.

It is possible to know the gene sequences which are involved in thermo tolerant enzymes by the genome of a thermophilic fungal species and to know the stability in different reaction temperatures. A study was done on hyperthermophilic archae *Pyrobaculum calidifontis* VA1 which produced a thermostable esterase (Hotta *et al.* 2002; Amo *et al.* 2002). The advantages of using microorganisms were found in the production of great variety of enzymes in a short period of time, survival in extreme environments of temperatures and/or acidic or alkali conditions producing natural, biodegradable compounds. There are many compounds reported in literature that are produced naturally by microorganisms which cannot be produced synthetically.

The biotransformation reactions depend on the quality of the substrate used. In case the substrate used is toxic, it might kill the microorganism. In that case the biotransformation reaction gets inhibited. In case the microorganism utilizes the substrate as an energy source than the product can be recovered. Biotransformation reactions come under the category of second generation process (Leresche and Meyer, 2006). Efficiency of microbial transformation can be improved by optimization of parameters for industrial products to be produced on large scale with greater yields. Kaur *et al.*, (2004) have studied the production, characterization and application of a thermostable polygalacturonase of a mould *Sporotrichum thermophils*. Kaur *et al.*, (2015) have discussed the solid substrate fermentation using agro industrial waste.

Cost plays a major role for the consumers. The products which are produced biotechnologically should have a reasonable price. This will help the purchasing power of the common man. The agro industrial waste materials can be used as the substrates. These generally contain polysaccharides, proteins and minerals (Sarma *et al.*, 2012) which are obtained from the plants. The feedstock can also be used for fermentation producing very high valued products. As these substrates are plant based they might improve the environment and also the disposal problem associated with agro-industrial wastes (Chapla *et al.* 2010). There are many agro-industrial wastes, such as sugarcane bagasse, sugar beet pulp, wheat stillage, apple pomace and cassava bagasse which have been used successfully for many products like ethanol, single cell protein and mushrooms (Sasaki *et al.* 1991; Pandey *et al.* 2000; Davis *et al.* 2005; Ruanglek *et al.* 2006 and Philippoussis 2009). Work has been done on the production of aroma/flavor compounds using agro-industrial wastes. Fruity flavor production by *Ceratocystis fimbriata* has been studied by Soares *et al.*, (2000). Scientists have also worked out on the application of coffee husk as the substrate to produce fruity flavour by solid-state fermentation. Renu Agrawal *et al.*, (1990) studied the different parameters affecting allicin formation as a secondary metabolites by static cultures of *Allium sativum*.

Pushpa Khanna *et al.*, (1980) studied the antimicrobial principles from tissue culture of some plant species. Renu Agrawal and Patwardhan (1994) studied the formation of piperine from *in vitro* cultures of *Piper nigrum* (paniyoor) L.

Soares *et al.*, (2000) have worked on the addition of glucose as additional carbon source as well as the effect of adding precursor and saline supplements on aroma production. They found acetaldehyde, ethanol, isopropanol and ethyl acetate as volatile compounds in the headspace. Compounds like ethyl isobutyrate, isobutyl acetate, isoamyl acetate, and ethyl-3-hexanoate were also present in low quantities (Soares *et al.*, 2000). Christen *et al.*, (1997) used wheat bran, sugarcane bagasse and cassava bagasse for aroma/flavor production by *Ceratocystis fimbriata* (Table 5 and 6 a, b).

Table 5: Utilization of wastes for aroma production at Large scale

Aroma/aromacompound	Waste-basedsubstrate	Microorganism	Productionscale	Reference
Pineapple aroma and banana odor	Coffee husk supplemented with glucose	*Ceratocystis fimbriata*	250 mL Erienmeyerflasks	Soares *et al.* 2000
Fruity aroma and banana aroma	Wheat bran, cassava bagasse, and sugarcane bagasse with supplements	*Ceratocystis fimbriata*		Christen *et al.* 1997
Vanillin	Waste residue of rice bran oil	*Aspergillus niger CGMCC0774 and Pycnoporus cinnabarinus CGMCCI115*	25 L fermenter	Zheng *et at.* 2007
Acetaldehyde, ethyl propio-nate. and 3-methyl butanol	Amaranth grain	*Rhizopus oryzae*	250 mL Erienmeyerflasks	Bramorski *et al.* 1998
Fruitv aroma compounds	Cassava bagasse and giant palm bran	*Kluyvemmyces marxianus*	250 mL Erienmeyerflasks	Medeiros *et al.* 2000
Acetoin	Molasses, soybean mealhy drolysate.andsupplements	*Bacillus subtil* is *C1CC 10025*	5 L fermenter	Xiao *et al.* 2007
Fruity aroma	Citric pulp, soya bran, sugarcane molasses, and other supplements	*Ceratocystis fimbriata*	250 mL Erienmeyer flasks	Rossi *et al.* 2009

Table 6a: Microorganisms producing value added Aroma Compounds from Wastes.

Microorganisms	Solid supports	References
Bacteria		
Amycolatopsis mediterranean MTCC 14	GOC and COC	Vastrad and Neelagund (2011a, b)
Xanthomonas campestris MTCC 2286	Potato peel	Vidhyalakshmi *et al.* (2012)
Pseudomonas spp. BUP6	GOC, COC, SOC, and CSC	Faisal *et al.* (2014)
Bacillus licheniformis MTCC 1483	Wheat straw, sugarcane bagasse, maize straw, and paddy straw	Kaur *et al.* (2015)
Fungi		
Aspergillus niger	Rice bran, wheat bran, black gram bran, GOC, and COC	Suganthi *et al.* (2011)
Aspergillus niger	Rice bran, wheat bran, black gram bran, and soybean	Kumar and Duhan (2011)
Streptomyces spp.	Household kitchen wastes	Ezejiofor *et al.* (2012)
Aspergillus oryzae	Soybean meal (waste)	Thakur *et al.* (2015)
Rhizopus arrhizus and Mucor subtillissimus	Corn cob cassava peel, soybeans, wheat bran, and citrus pulp	Nascimento *et al.* (2015)
Aspergillus niger	Rice bran, wheat bran, black gram bran, GOC, and COC	Mahalakshmi and Jayalakshmi (2016)
Aspergillus terreus	Palm oil cake	Rahman *et al.* (2016)

Source: Sadh *et al.* Bioresour. Bioprocess. (2018) 5:1-15.

According to the authors, sensory characteristics of the final product could vary with respect to the synthetic supplement added to the process and the aroma production process was found to be growth associated. Nearly 20 compounds including alcohols, aldehydes, esters and ketones were detected in the headspace gas mixture of the process (Christen *et al.*, 1997). Rossi *et al.*, (2009) found the utilization of citric pulp (waste generated during citric juice production) by microorganism *Ceratocystis fimbriata* for aroma compound production. They found that the supplemention of carbon and nitrogen sources played a major role while using the waste feedstock to improve aroma production. Use of supplementary nutrients like soy molasses, soy bran, sugarcane molasses and urea gave good results. It was found that using citrus pulp with soy bran (50 %), sugarcane molasses (25 %) and mineral saline solution had positive effect on the aroma production (Rossi *et al.*, 2009). Han *et al.*, (1998) studied the biotransformation of lavandulol by fungus *Glomeralla cingulate*.

Vanillin production using ferulic acid from waste residue of rice bran oil has been worked out by Zheng *et al.*, (2007) in a two-step process. During this, initially vanillic acid is produced by *Aspergillus niger* CGMCC0774 which is later bioconverted into vanillin by *Pycnoporus cinnabarinus* CGMCC1115 (Zheng *et al.*, 2007). Edible fungus *Rhizopus oryzae* is known to produce volatile aroma compounds such as acetaldehyde from solid agro-industrial waste. Bramorski *et al.*, (1998) worked on the production of volatile metabolites by *R.oryzae* using different tropical agro-industrial waste substrates, such as amaranth grain and cassava bagasse (Bramorski *et al.*, 1998)by a solid-state fermentation method. Lot of work is being carried out by using yeast strains. Medeiros *et al.*, (2000) have studied aroma compound production from different agro-industrial wastes by the yeast *Kluyveromyces marxianus*. They found that substrates like cassava bagasse and giant palm bran had the ability for high bioconversion. They also worked on the addition of glucose as an additional carbon source in the reaction. This resulted in producing higher yields of aroma compounds. Medeiros *et al.*, (2000) observed that in the headspace alcohols, aldehydes and esters were found. The fruity aroma was imparted by the esters. Looking into the literature it is seen that among the various organisms which use agro-industrial wastes it is the fungus which is mostly used for biotechnological production of aroma compounds. Recently, much work has been carried out by yeast strains and they seem to have the potential. Many scientists are also looking into the probability of using bacteria for aroma compound production from agroindustrial by-products. Acetoin production by *Bacillus subtilis* CICC 10025 has been worked out by Xiao *et al.*, (2007) using molasses and soybean meal hydrolysate. In this study it was also found out that molasses and soybean meal hydrolysate helped acetoin production with (99 %) significant level (Xiao *et al.*, 2007). There is a need to screen crude extracts for biochemical compound production. It is also important to know the structure of the active principles for further studies. This will further help in metabolic pathways and mechanism of action of bioactive compounds.

Table 6b: Effect of Phenolic compounds produced by Agriculture Wastes

Phenolic compound	Effect	References
Anthocyanin	Improvement in visual capacity	Lee *et al.*, (2005)
	Improvement in brain function	Kang *et al.*, (2006); Shin *et al.*, (2006)
	Prevention of diabetes and obesity	Tsuda *et al.*, (2003; Guo *et al.* (2007)
	Decrease in serum triglycerides total cholesterol	Xia *et al.* (2006)
	Decrease in lipid hydroperoxides	McAnulty *et al.*, (2005)
	Decrease in blood pressure	Herrera-Arellano *et al.*, (2007)
Ellagic acid	Slow down tumour growth protection	Harris *et al.* (2001)
	against cardiovascular diseases	Aviram *et al.*, (2000)
	Anticancer	Aggarwal and Shishodia (2006)
Isoflavones (phytoestrogens)	Prevention of osteoporosis	Williamson-Hughes *et al.*
	Lowering effect on LDL and cholesterol	(2006) Jayagopal *et al.*
	Improvement of cognitive function	(2002) Lee *et al.* (2005)
	Reduction of colon cancer	Cotterchio *et al.* (2006)
	Modulation of immune function	Rayan-Borchers *et al.*, (2006)
Resveratrol (3,5,4-trans-trihydroxystilbene)	Cardioprotective effect	Bertelli *et al.* (1995)
	Anticancer	Jang *et al.* (1997)
	Prevention of Ischemic damage	Bradamante *et al.* (2004)
Flavonones (naringenin and hesperitin and corresponding glycosides)	Cardioprotective effect	Jeon *et al.* (2007)
	Decrease in liver toxicity	Kaur *et al.* (2006)
	Protection against arthritis	Kawaguchi *et al.*, (2006)
	Decrease in cholesterol	Williamson and Manach (2005)
Proanthocyanidins	Decrease in blood pressure	
	Maintenance of endothelial function	
	Improvement of blood circulation in legs	Christie *et al.* (2004)

Use of genomics, proteomics and metabolomics to obtain bioactive compounds from agro-processing wastes will be effective. Agricultural and industrial residues are attractive sources of natural antioxidants (Volf and Popa 2004). Among the agri wastes which are found in large quantities are lignocellulosic waste which comprises of rice straw, wheat straw, sugarcane bagasse, sweet sorghum bagasse, cotton stalk, etc. In South East Asian countries rice straw is burnt after crop harvest in the fields. This causes loss of biomass and environmental pollution. Bioactive compounds are extra nutritional constituents that occur naturally in small quantities in plants and food products (Kris-Etherton *et al.*

2002). These bioactive compounds are present in different parts of the plant like leaves, stems, roots, tubers, buds, fruits, vegetables, seeds, peels, flowers and plant-derived foods and drinks (tea, coffee, alcoholic beverages). They are required for the growth, maintenance and repair of the body. Among the common bioactive secondary metabolites are antibiotics, mycotoxins, alkaloids, food grade pigments, plant growth factors and phenolic compounds (Holker *et al.*, 2004; Kris-Etherton *et al.*, 2002). The method of extraction is given in Figure-3. Recently, many food bioactive constituents are in the markets which are being utilized in pharmaceutical products especially phenolic compounds in the form of pills, capsules, solutions, gels, liquors, powders and granules. These affect beneficially in physiological functions. These products are termed as "nutraceuticals" as they provide nutrients and are used in pharmaceuticals. These enhance health that can not be obtained from normal foods (Zeisel, 1999). As the public is becoming very health-conscious these types of health-promoting products are becoming popular. Carotenoids like lycopene, alliaceae (garlic, onion) extracts which contain sulphur derivatives (alliin and allicin), glucosinolate extracts and phytosterol extracts,carotenoids, polyphenols such as anthocyanins, proanthocyanidins, flavonols stilbenes, hydroxycinnamates, coumarins, ellagic acid (EA) and ellagitannins (ETs), isoflavones and lignans are available. Prabha *et al.*, (1991) studied the intermediary metabolism and thymol formation in *in vitro* cultures of *Carum copticum*.

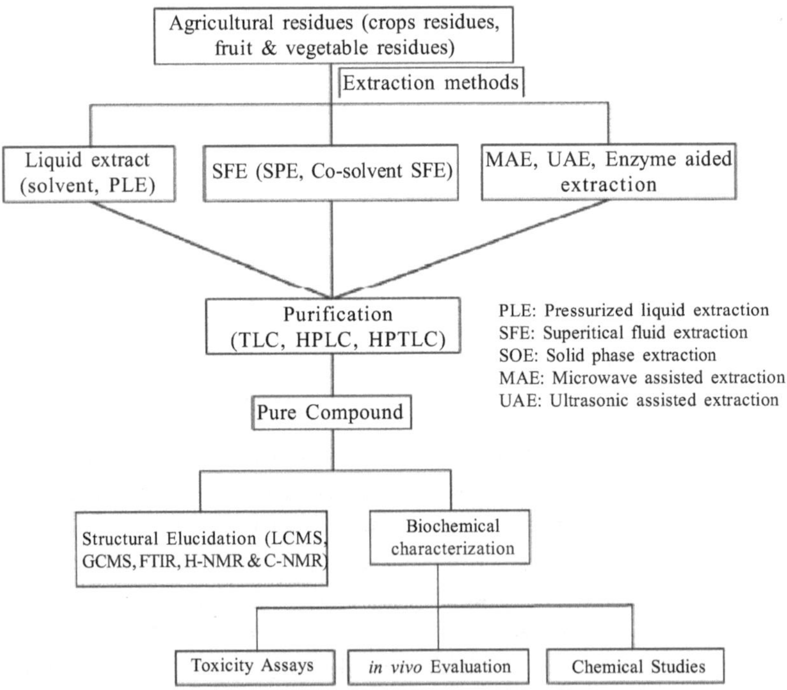

Fig. 3: Schematic diagram for extraction, isolation and characterization of bioactive compounds from agricultural residues.

5

Novel Approaches for Natural Product Biogeneration /Immobilization Techniques, Genetic Engineering

The possibility of making genetic changes in the laboratory in plants, animals and microbes used in food started almost 10,000 years ago. Along with natural changes taken place during the evolution totally different species have come into existence. This has helped in increased food production with better taste and nutrition. The losses caused by biotic and abiotic stresses from fungi and bacterial pathogens have reduced.

Food scientists are engaged in designing new genetic modifications which will help in the identification, selection and analyses of such plants and organisms. According to UPOV (2002) any new variety must be genetically different, must be genetically uniform throughout the population and should be genetically stable. To meet all these requirements a plant species takes about 12 years. A new variety must be superior and beneficial for consumption. They must also have environmental fitness. Sometimes it is observed that some traits that enhance plant resistance to disease can be harmful to human beings.

5. a. Strain Improvement

Interspecies Crossing

By various means interspecies crossing takes place among closely related species. This has been seen in cultivated *Avena sativa* and its weedy relative wild oat *Avena fatua.* Cross-pollination helps in the exchange of genetic information. Sometimes there is a possibility of genes from one species to integrate into the genomes of distant species. Together, both by natural and human intervention it may give rise to a high yielding variety. It has been observed that some of the wheat varieties carry genes from rye. It is also possible for common potato, *Solanum tuberosum* to be crossed with other species like *S. acaule* (Kozukue *et al.*, 1999) or *S. chacoense* (Sanford *et al.*, 1998; Zimnoch-Guzowska *et al.*, 2000) to produce high yielding strains. In Chromosome engineering portions of chromosomes from near or distant species get recombined by a natural process called chromosomal translocation. This is a nonrecombinant deoxyribonucleic acid (rDNA) procedure. Sears (1956, 1981) was successful in transferring some valuable traits like pest or disease resistance which were not possible otherwise into crop species. The technique was found to be limited as during transferring large segments of chromosomes a number of neutral/ detrimental genes also went along with it. Mutation breeding is another method to induce genetic changes. This involves exposing plants or seeds to mutagenic agents like ionizing radiation or chemical mutagens like ethyl methane sulfonate. These agents induce random changes in the DNA sequence. When the breeder is using at that time the dose of the mutagen it can be adjusted to result in non lethal mutations. Looking into the literature this method produces a large number of mutagenized plants/seeds. These are grown upto reproductive maturity giving rise to mutant progeny. These are tested for phenotypic and chemical changes. Worldwide, more than 2,300 different crop varieties have been developed using induced mutagenesis (FAO/IAEA, 2001) and about half of these have been developed during the past 15 years. The United States has come out of mutated new crop varieties of wheat and grapefruit. However, as the work was started way back in the last century there are no molecular characterizations in records.

5. b. Genetic Improvement

Presently, chromosome engineering is becoming more competitive due to the refinements that allow the plant breeders to restrict the transferred genetic material. This is seen to focus on the gene of interest (Lukaszewski, 2004). Thus rDNA technology has the ability to transfer only small pieces of DNA. This has been tried on many crops like corn, soybean, rice, barley and potato (Gupta and Tsuchiya, 1991). The role of genetically modified (GM) crops for food security is the subject of public controversy. GM crops could contribute to food production increases and higher food availability. There may also be impacts on food quality and nutrient composition.

The GM crops may have a great influence on the farmers' in many ways. It might affect their income and they may have to bear losses. As such the small farmers are highly undernourished all over the globe. If we look into the security impact of the GM crops in India where insect-resistant GM cotton has been widely adopted it is found that the adoption of GM cotton has significantly improved calorie consumption and dietary quality, resulting from increased family incomes. The food security has infact increased by 15–20% among cotton growers (Qaim and Kouser, 2013).

5. c. Classical Genetic methods

i. Mutation/ random selection

Some plants cross-pollinate resulting in a fertilized hybrid embryo which does not mature and sprout. Modern plant breeders have first pollinated two species naturally and then removed the plant embryo while growing. This was placed in a tissue-culture environment for development. This procedure of embryo rescue is an intermediary step for transferring genes from distant, sexually incompatible relatives by compatible relatives of both the donor and recipient species. This does not come under the category of genetic engineering. The procedure helps in producing new varieties.

ii. Protoplast Fusion

Recent developments in scientific techniques have developed recombinant genes from different plant sources. In the procedure of somatic hybridization two cells are fused. The cell wall is removed from the cells while growing in a culture medium. The use of enzymes like pectinase, cellulase and hemicellulase has been helpful. These cells without the cell wall are known as protoplasts. Protoplast fusion can be done by electric shock also. When the two protoplasts fuse, the resulting somatic hybrid contains the genetic material from the two sources. This helps to overcome the physical barriers. After the fusion, if the fusant shows compatibility then a new cell wall grows around it. This then divides mitotically ultimately growing into a hybrid plant with the genetic characteristics of both the parents. Protoplast fusions are obtained easily but many of them do not have the ability to regenerate into a whole organism. Smita Rao *et al.* (2003) studied the intra specific auxotrophic hybridization in *Aspergillus niger* and *Penicillium digitatum* by protoplast fusion for strain improvement in verbenol production. Vidya and Renu Agrawal (2003) studied the production of verbenol, a high valued food flavourant from improved culture strain of *A.niger* using protoplast fusion. The cultures utilised were locally isolated from the waste soil where lime trees were growing.

Agrawal *et al.* (1999) studied the Strain improvement of *Aspergillus* sp. and *Penicillium* sp. by induced mutation for biotransformation of alpha-pinene to verbenol. Variants of *Aspergillus* sp. and *Penicillium* sp. obtained after treatment with colchicine, ethyl methane sulphonate (EMS), or ultraviolet (UV) irradiation indicated varying levels of significant increases in their efficiency to transform alpha-pinene to verbenol. In case of *Aspergillus* sp, the UV-induced variant was the best performer with a 15-fold increase in biotransformation efficiency compared to the wild type. In case of colchicine and EMS-induced variants the biotransformation increases were 2- and 8-fold, respectively. The UV-induced variant of *Penicillium* sp. was capable of eight fold increase in the efficiency while the colchicine- and EMS-induced variants were 1.5 and 2-fold, respectively. The variants were characterised with respect to changes in colony morphology, spore dimension, DNA content, and products formed, *viz*. verbenol and verbenone.

Swanson *et al*., (1988) studied the characterization of herbicide tolerant plants in *Brassica napus* after *in vitro* selection of microspores and protoplasts. In plant breeding this non-genetic engineering technique is not used often. When the plant cells are grown *in vitro,* most of the times spontaneous mutations take place. These are known as somaclonal variation. Newer varieties have been obtained by using the technique of tissue culture. It was the work of two scientists namely, Larkin and Scowcroft (1981) who realized that soma clonal variation can give rise to beneficial genetic variability. Among these substances it is one or more unusual nonprotein amino acid called opine which can be translocated in the plant. Many strains of *Agrobacterium* lack disease causing genes and help in the transfer of DNA.

iii. Molecular Genetic Improvements

Microprojectile Bombardment

It was in 1987 when scientists, Klein *et al*., (1987) found out that it was possible to deliver naked DNA to the plant cells when microscopic pellets of DNA are shooted on them. This has been found successful in corn, rice and some other cereal grains. Initially micro projectile delivery was utilized for the production of genetically engineered plants on large scale.

Electroporation: In the process, the protoplasts take up macromolecules from their surrounding fluid which is facilitated by an electrical impulse. The cell walls are removed from the cells when they are growing in a culture medium to form protoplasts. To these protoplasts when a known DNA is supplied under an electrical current it destabilizes the cell membrane temporarily and allows the DNA to enter the cell. Later these transformed cells regenerate their cell walls. They grow into whole, fertile transgenic plants.

Microinjection: DNA can be directly injected into the cells. The cells with new DNA survive in many cases and the cell DNA integrates with the injected DNA. However, the process is labor intensive and inefficient.

Transposons/Transposable Elements: Transposons are short, naturally occurring pieces of DNA. These have the ability to move from one location to another in the genome. Such genes are present in many plants and some animal (insects and fish) species. The presence of such genes was first described by Barbara McClintock in corn plants during the 1950's (Cold Spring Harbor

Laboratory, 1951). To study mutagenesis and the mechanics of DNA recombination such transposons have been investigated. Work is underway to improve commercial crops. Mohimani *et al.*, (2014) studied the automated genome mining of ribosomal peptides.

Nontransgenic Molecular Methods of Manipulation: In this method new genetic characteristics are put into the plants and animals without actually inserting them into the recipient's native genome. DNA of interest may be delivered to a plant cell where it expresses a new protein which gives rise to a new trait. This happens without actually getting integrated into the host cell DNA. It has been possible to modify virus strains that can carry genetic material into a plant cell, replicate and thrive without integrating into the host genome. Though there is a disadvantage without integration it is possible that new genetic material might get lost during meiosis. This does not allow the seed progeny to express the new trait.

Growing by vegetative means like grafting/cuttings are done in food plants. The new genes enter in a nonsexual manner and are not the products of rDNA as there is no recombination or insertion of introduced DNA into the host genome. Even then they have a new DNA and new traits. According to McHughen (2000) no such products are in the market in the United States or any other country. If one looks into the modern breeds of livestock they are very different from the earlier ones due to the breeding strategies. In an example of cow the milk production per cow has increased among Holstein dairy cattle. Similarly, breeding programs have resulted in fast-growing pigs (Notter, 1999). Due to genetic selection the egg production in chicken has increased (Ayyagari *et al.* 1980). In animal agriculture the emerging biotechnologies include assisted reproductive technologies (use of naturally occurring hormones, such as recombinant bovine somatotropin, marker-assisted selection, biotechnologies to enhance reproductive efficiency without affecting the genome) and biotechnologies to enhance expression of desirable genes.

Assisted Reproductive Procedures

Frozen semen for artificial insemination (AI), along with sire testing and sire selection has resulted in different breeds of livestocks. This has changed the genetic quality of the livestock, especially dairy cattle. In an observation made by Majeskie (1996) the milk obtained from Holstein cows increased around three times when a combination of semen from select bulls was used. In another study, Diamond (1999) and Hale (1969) studied the improved milk production management. Using sophisticated statistical models to predict breeding values, sire testing and selection, cross breeding and marker-assisted selection, along with AI, have greatly advanced the production characteristics of livestock. It is expected that AI will continue to be an integral tool in animal production systems.

Overall breeding and system for selection to propagate desired genotypes in animals with desirable traits can be obtained by various methods as:

Embryo Recovery and Transfer and Super ovulation. By this method the animals are able to produce more off springs into the gene pool (Seidel, 1984). Embryos those are frozen and stored before being used to initiate a pregnancy result in 40,000 to 50,000 beef calves per year (NAAB, 2000). The sex of the semen and embryos can be controlled by these emerging technologies. According to Johnson (2000) single-sex sperm can be sorted out as X and Y sperm which will benefit the livestock industries to a large extent.

***in Vitro* Maturation and Fertilization of Oocytes**: Techniques for recovering and maturing immature eggs or oocytes can produce many thousands of embryos in one day in a medium with hormones. This can be fertilized with the live sperm. *In vitro* culturing of the resulting zygotes upto the blastocyst stage helps before transferring to the recipient females (First, 1991). This procedure has given rise to almost 4,000 calves in a single year (NAAB, 2000).

Embryo Splitting: According to Heyman *et al*., (1998) splitting or bisecting the embryos gives rise to zygotic twins which are non-GE clones. Therefore, they become genetically identical as they have the same nuclear and mitochondrial genes. The variation in paternal twins is less as it has only one X chromosome where as it is found more in maternal twins. There may be differences in the distribution of mitochondrial DNA which may affect the phenotype. As per a study (NAAB, 2003) when these embryos are put in an empty zona pellucida and transferred to recipient females, which carry them to term variation was noted. Through 2001, a total of 2,226 registered Holstein clones754 males and 1,472 females were found. These were produced from embryo splitting. It was found that it produced 1 to 2 percent of calves.

Genetic Engineering: In a detail report on animal biotechnology (NRC, 2002) has given various techniques of cloning and its implications for predicting and assessing adverse health effects that may be associated with this technique. They have also discussed different techniques employed to introduce novel genes into domestic animals. Varoius transgenic approaches applicable to animals include accessing the germline of animals. Germline refers to the lineage of cells that can be genetically traced from parent to offspring. It is possible to access the germline of animals using one of five methods (NRC, 2002). It directly manipulates the fertilized egg after it has been implanted in the uterus, manipulating the sperm that produces the zygote, manipulating early embryonic tissue in place, using embryonic stem cell lines in early embryos and manipulating cultured somatic cells to transfer their nuclei into enucleated oocytes.

Transfection: These are done by microinjection of DNA into the nucleus of anchored cells or electroporation. In the process the DNA is introduced through the pores in the cell membrane by pulsed electrical charges. Then the cell membrane is neutralized and the DNA is introduced to improve the passive uptake. In lipofection the DNA and sperm-mediated transfection is usually used in conjunction with intracytoplasmic sperm injection or electroporation. In animals microinjection is very inefficient means of creating transgenic animals. According to a study done by Rexroad (1994) microinjection in livestock embryos yields a very small percentage of transgenic animals. Moreover, successfully microinjected transgenic animals do not necessarily pass their transgenes on to their offspring (NRC, 2002).

Retroviral Vectors: In this method virus strains are modified to carry genetic material into a cell. It differs as in this method after the novel DNA is delivered the viral replication process integrates itself into the host genome.

Knock-In and Knock-Out Technology: In this technology organisms are created that lack specific genes or exchange one of the existing gene by another that has been engineered. The addition ("knock-in") or deletion ("knock-out") of specific gene by mutations or genetic engineering based on homologous recombination is being worked out in mice. Presently this technology is not found to be efficient and practical for domestic animals. However, NRC (2002) has shown its use in domestic sheep and pigs.

Marker-Assisted Selection: In this method specific genetic markers are coupled to a particular trait wherein a link is established for a desirable trait which needs to be inherited. For studying complex traits this method plays a major role in animal breeding and selection strategies as it is governed by many genes (Georges, 2001). The method helps to identify useful segregated markers that are economically important. Markers that accompany multiple genes controlling more complex traits include meat tenderness, taste, growth, offspring size and disease resistance which can be dealt with. This will improve the health of the animals and other traits (Dekkers and Hospital, 2002).

Freking *et al.*, (2002) have worked on sheep. They have worked on the Booroola gene in which a single-nucleotide base change is responsible for the callipyge muscle hypertrophy phenotype. Galloway *et al.*, (2000) have shown the mutations in an oocyte-derived growth factor gene (BMP15) which causes increased ovulation rate and infertility in a dosage sensitive manner. Animals that are important to agriculture are studied for sequencing of the genomes to identify genes that may influence reproductive efficiency. Fletcher, (2003) has shown a growth-hormone receptor variant on bovine chromosome. This may enhance the yield and composition of milk and decrease milk fat from 4.4 percent to 3.4 percent.

Nontransgenic Methods of Animal Manipulation

For better reproduction, lactation and growth in domestic animals endocrine functions can be modified. In a study done by Draghia-Akli *et al.* (2002) in pigs and rats it was found that the hypothalamic-specific expression of growth hormone is not essential since ectopic expression of a cloned DNA. According to Khan *et al.* (2002) a neuropeptide can be genetically driven by a synthetic muscle-specific transcriptional promoter to elicit increases in both growth hormone and insulin-like growth factor-I. With the use of specific hormones and growth factors during critical developmental periods it is possible to increase the uterine capacity and also milk production using biotechnological methods.

Genetic Modification of Microbes

Genetically modified (GM) microbes have become popular for improving production of food. The quality of products like wine, bread and cheese depend on microbial starter cultures. Bacteria and yeasts have different genetic traits. This helps in developing useful microbial strains through simple selection or induced mutation. An important role is played by these microorganisms towards food production. They produce enzymes and other metabolites used in food production and processing and also for food spoilage. These starter cultures are responsible for fermentation. In fermentation industries the fermented products are prepared utilizing starter cultures. The very common microorganisms are lactic acid bacteria (LAB) and yeasts (*Sacchromyces cerevisiae*). These cultures can also be modified using genetic modification methods. These include selection, mutagenesis, conjugation and protoplast fusion.

Earlier to genetic methods other methods like chemical and ultraviolet rays were used which induced mutagenesis. After this procedure an enrichment or selection process came into being for mutants with superior characteristics. Apart from this traditionally, conjugation method has been commonly used where genetic exchange takes place. Conjugation can occur between LAB strains as well as between LAB and other bacteria (Steenson and Klaenhammer, 1987). As the changes take place at the genetic level they are classified under GE organisms. Protoplast fusion is another method where recombination of two strains with desirable quality takes place. Protoplast fusion has been used as a mapping method in bacteria and successfully produces strains of LAB with desired characteristics (Patnaik *et al.*, 2002). By this method it has been possible to generate yeast strains that can produce greater number of biochemical substrates which are utilized in the fermentation process (Pina *et al.* 1986).

Today, the industries are looking for starter cultures that have desirable traits with higher efficiency after being engineered. Transformation is another method where by the DNA of interest is introduced directly into recipient cells by making them permeable with the use of either of chemical agents, enzymes or electroporation. According to Kondo and McKay (1984) the first method developed for LAB was plasmid protoplast fusion in which recipient cells are stripped off their walls and subsequently fused with polyethylene glycol which traps the newly introduced DNA between the cells. Electroporation is another method which was developed during the late 1980's especially for LAB. In this method electrical currents are given that creates pores in the cell envelope. This allows DNA from other sources to enter (Luchansky *et al.*,1988). However, a lot of work is presently being carried out to look into the efficiency of such cell fusions.

Recombinant DNA also can be introduced into LAB using a technique called transduction, in which a bacteriophage is used to move DNA from one strain into another (Bierkland and Holo, 1993). In transduction many deletions take place within the plasmid (known as transductional shortening that are typically of undefined length). Transformations are simpler and efficient in microbes than in higher organisms. Studies on microbes have shown as to how the analogous events take place in molecular genetics of higher organisms.

Application of new enrichment and separation technologies for natural product research was discussed by Günther (2016) in 10th international symposium on chromatography of natural products for the application of analytical methods in the development of natural products.

Novel enrichment and separation technologies have become very important as they are application oriented for the extraction of natural products. Novel new advancements in chromatography have explored areas of natural product isolation. Special knowledge and technologies are necessary to perform analysis of plant ingredients because usually they have hundreds or thousands of small low molecular weight ingredients in different concentrations. The achievements in natural product research are largely based on the constant development of highly selective and sensitive analytical technologies (Zhu *et al.* 2015). Novel enrichment and purification methods based on modern solid phase extraction (SPE) technologies are applied to reduce the complexity of plant-extracts, while μ-HPLC is used for separation, pre-concentration and fractionation. The opportunity to hyphenate these techniques to robotic systems permits high-through put screening. Significant progress has been made in the development of novel stationary phases which can be tailored to a specific application allowing endless possibilities in terms of selectivity tuning (Majors, 2001). High-resolution mass spectrometry facilitates the identification and quantitation of active components in natural products. An alternative offline coupling to MALDI-

TOF/MS allows easy determination of active compounds using novel matrix-free approaches (Lucinda *et al.*, 2002).

In phytopharmacy and food analysis there is a combination of many separation techniques along with spectro scopic methods. Near and mid infrared (NIR and MIR) spectroscopy enables very fast and simultaneous qualitative and quantitative analysis of raw plants and liquid extracts where the destruction is minimal (Cozzolino, 2015). The distribution of active ingredients in plant materials can be studied by infrared imaging which has a resolution upto 5 μm. These have applications in medicine, phytopharmacy, phytocosmetics and also in nutrition science (Stecher *et al.*, 2011; Koeck *et al.*, 2014).

6

Future Implications and Prospects

Commercial Utilization of the Products Produced

In the Asian and Pacific 'region, it is China alone which produces the largest quantities of agricultural waste and also crop residues. This is followed by India. From the production of rice, corn and wheat alone 587 million tons of residues are generated per year in China. Maheshwari (2014) has surveyed the proportions of waste that Malaysia generates from the production of rice, palm oil, rubber, coconut and forest products. In Myanmar the crop waste and residues constitute about 4 million tons per year out of which more than half constitutes rice husk. The annual animal waste production is about 28 million tons. Maximum is from cattle husbandry. Commercial rice milling generates around 2 million tons of paddy husk per annum where as coir (the fibres from coconut husks) processing generates 70,0000 tons of coir dust per year (Maheshwari, 2014). In case of Thailand about 4.6 million tons of paddy husk, 35 million tons of rice straw, 7 million tons of bagasse and more than 25 million tons of animal wastes are produced (Maheshwari, 2014). Other countries such as Australia, Cambodia, Japan, Lao People's Democratic Republic, Nepal, New Zealand, Republic of Korea, Vietnam and Small Island States in the South Pacific also generate huge quantities of agricultural wastes and residues (Maheshwari, 2014). During various surveys it has been observed that most hazardous wastes are the by-products from industrial, agricultural and manufacturing processes, nuclear establishments, hospitals and health-care facilities. Primarily, high volume generators of industrial hazardous waste include the chemicals, petrochemicals, petroleum products, metals, wood treatment, pulp and paper, leather and textiles. Angerosa *et al.*, (2012) have written a book on animal wastes, water quality and human health. It gives an insight towards the utilization of these wastes. It also includes energy production from plants which include coal-fired and nuclear power plants. Hazardous wastes are generated by small and medium sized industries which include auto and equipment repair shops, electroplating, metal finishing shops, textile factories, hospitals, healthcare centres, drycleaners and pesticide users. In the Asian and Pacific regions the waste includes solvents, chlorine bearing waste and pesticide or gano phosphate herbicide, urea and fungicide bearing waste.

With the growing population the farmers must produce more food per unit of land, water and agrochemicals. However, they will be facing many challenges like that of climate change, volatility, shifting nutrition needs and scarcity of the physical factors of production. There is a paradigm shift taking place in agriculture. If one looks into the most important priorities for sustainable food and agriculture systems in the next 20 years then farming has enormous impact on the world's most critical resources. The resources versus needs have to be looked into so that enough food is provided to every body. Small and medium farmers need to work together in a progressive manner. The focus needs to be on the use of natural resources in a sustainable manner with better production and better food systems. Newer and innovative approaches need to be undertaken with checking the environmental impacts and simultaneously looking into the social consequences in evolving agri-food systems. It has been seen that in developing countries large amount of food produced does not reach the market each year. Also in developed countries much food is wasted. In a statistical study it has been found that the obese/malnutrited people are almost equal in number. Therefore, from growing more food the attention should be on more nutrition or healthier food. The food resources are being used in an inefficient manner bringing crises and volatile market dynamics.

Today, due to the shift in governance concentration in supply gives a major risk. Environment is playing a major role in agriculture today. The limitations of natural resources needs to be studied properly so that ultimately improved social out comes are achieved. These newer and innovative concepts in agriculture are being taken up by many food companies rather than international agencies or multilateral and bilateral institutions. The dynamics of local and global food systems are changing these days due to the upcoming corporate houses. It will be important that public sector and private sector investment policies integrate among themselves and have a common goal. Agriculture is one of the major occupations of people however it is also a threat to a healthy environment. Much of the lands are lost every year due to soil erosion especially in the developing countries. It is important that agriculture gets integrated to the ecosystem management which will help in development. We need to look for achievable solutions. Recently in many countries the governments have retreated their investment from agriculture and are changing from public to private. This is because private enterprise is powerful and has increased concentration all along the supply chain. Firms are also facing new governance challenges with volatility in supplies and markets. The systems are becoming more stringent and transparent. Forward-looking firms recognize the need to create shared value and not just profit if they have to thrive. The governance by the public work should benefit both public and private sectors. The improved methods must have the new learning methods for producers, industries and policy makers. The public sector must be people friendly. Today the major areas to be focused are the organized small and medium farmers, women farmers, human nutrition and methods to have high produce yields along with clean environment. Newer and innovative technologies should be suitable in any socioeconomic and ecological conditions. Wastage should be minimized along the entire food chain. Decentralized biofuel systems need to be worked out to promote energy. The output should be reviewed quarterly or once in six months. It is important that we identify the waste and the industries utilizing with transparency so that a system can be made for public interest.

Motivate and reward investments and business systems that have given good results.

Parts of peri-urban China provide diversified subsistence for the household by having good yields of food crops in small areas. It is observed that Chinese produce them without the use of chemicals. These areas provide the food produce to the urban region of China. This helps both the rural and urban population along with creating green environment. This policy needs to be followed by other countries. Apart from this the availability of raw materials, seeds, post-harvest technologies of the produce, shelf life of the produce and processing needs to be taken into account for the profit to the farmers.

When farmer organizations and extension services are well coordinated and connected it turns very helpful. The public policies and institutions for any food technology should be people friendly. For diversifying or making new food products public-private models need to be effective. Farming women play a very vital role. Infact they are the ones who take care of the farms and harvesting to a large extent. Therefore, it is important to empower them for food security. Agricultural programs and policies must address women. There are many agriculture organizations, like IFAD, FAO, WFP, the World Bank and the United Nations Committee on World Food Security which follows gender equality as a pillar of development. The time given to them as land tenure is very vital as it will enable them to farm with surety and invest in their farms and soils. Awareness camps for more information and training may definitely help.

Private companies can help to a large extent as they benefit public. They can give incentives for their good activities which may help in bringing better results.

Many a times the focus goes on high yields rather than better nutrition. We need to focus our goals towards good and healthier nutrition. In recent times problems like obesity and malnutrition have cropped up even in highly developed nations. High nutrition could be supplied through mid day meals in schools and local markets via public funding. It should be of low-cost which could at the same time provide high-nutrition. Much work is also being undertaken to enhance the shelf life of food products. Recently, scientists have come out with innovative and newer food products which could take care of diseases like cardiac, cancer, diabetes, malnutrition, parkinson, alzeimers and obesity. Now time has come when agriculture will take a central place in the management of healthy ecosystems. The fact is that we have come out of 'yield per hectare' towards 'productivity' in agriculture that values ecosystem services like that of water infiltration, carbon sequestration and conservation of biodiversity.

It is important to check the soil health. When the soil is healthy definitely we will have higher yields of produce. Natural resources like water, genetically engineered crops, organisms and also mechanical or drip irrigation may increase the yields of the crops. The health of the soil is also interlinked with water management. The role of water in increasing productivity is very important. Today, it is observed that the quality of water is deteriorating in many parts of the world. Many fresh water sources are declining which is negatively impacting the agriculture. By rehabilitating agricultural areas that have got eroded and degraded can be conserved.

Adaptation of climate-smart production systems will be vital in future. Agriculture contributes to climate change (GHG emissions mostly via livestock, deforestation and fertilizers) which in turn affects the shift in climate. If any agricultural area is impacted it will have to rely on food from healthy regional and international markets. Parameters for food production have to become resilient. For improving indigenous crops there are many opportunities, many of which have already adapted to harsh environments and resistant to diseases. Even where agricultural conditions could benefit from climate change (i.e. more rain in semi-arid areas or higher temperatures in cold regions) the near-term benefits are still likely to be very limited as farmers adapt to new conditions, soils and cultivation methods. This will require considerable investment in more adaptable crop varietals and expanding farmer training in appropriate methods.

There is a need to combine market mechanisms and policies which will help in taking agriculture to greater heights. The systems must improve the shelf life of the products and improve the delivery of ecosystem services. Public goods like biodiversity, genetic resources and traditional knowledge should be transparent so that they can be utilized by all. There is a need for cross-border cooperation technology which is affordable and can be scaled up. It has been observed that many times projects stay on paper and the objectives are not fulfilled. Therefore if the technologies are of low cost and are available to small and medium producers it will be helpful in setting up business which will ultimately be for national gain. If the wastes are significantly reduced along the entire food chain and utilized for the production of products than it will help the nation and the world. This will also add a considerable percentage to our food supply and will have positive impact on the environment. Production of biofuel can help in achieving security for energy and livelihood. The potential of biofuels to provide energy, security and rural livelihood is known and is very helpful. However, it must be taken into account that it does not negatively impact the food economy.

The results should be measurable and transparent. It is difficult to manage if it is not measurable. Industries must invest their money in projects which have surety of performance. There should also be set policies in the system. This will bring accountability and hike in work ethics. Recent literature gives clear understanding of what works and what does not. Time has come when assessment of economic or financial outcomes can be done along with social and environmental ones as well. When the outcomes can be managed then it is easier to set the learning pathways, smart investments and policies towards the approaches which will provide effective multi-dimensional solutions. The effects of subsidies, green incentives and environmental taxes can also be set properly.

The private and government institutions should have a vision with a purpose. This will allow the parties to invest in community well-being and to secure and restore natural resources in farming especially water, grasslands and forests. This in turn will increase the agricultural productivity and enhance resilience to climate change. If the policies are set right then it will be possible to bring government, civil society and private finance investors to come together for the improvement in agriculture.

Any policy must value both societal costs (carbon emissions, ecosystem degradation and ground water depletion) and societal benefits (crop genetic diversity, carbon sequestration, water purification and nutritional value of foods). For sustainable agriculture and food systems it is important to have a combination of clear policy and price signals. The benefits to society will be an efficient way towards achievements.

Innovations should be such which could promote agriculture effectively. It may be possible that in better soil and watershed protection systems the public sector might effectively support private participation. If it is possible to create a strong policy and incentives to reduce the considerable waste the distribution system of the food may become better.

Presently, the relative investment in agriculture has been decreased by the Government. This would in turn increase investment in private sector and thus the efficiencies in some areas. However, it is important to put emphasis on handling the issues which can improve the conditions of poor, decrease poverty like food security, wide availability of inputs, extension and knowledge services, local market and storage systems and increase in land tenure. If investments and businesses are rewarded it will definitely bring public good.

New institutions with better standard systems are collaborating with business partners and producers in order to accelerate towards sustainability. With increasing transparency smart firms are recognizing the limitations of various resources from crops to goodwill. For good standards an important aspect is food safety. This can be accompalished with a combined effort of public and private investment along with clear governance. Voluntary standards, including those involved in carbon markets, social accounting, organic and other environmental standards are among the new market mechanisms that connect profits with the provision of public benefits. It may not be possible for many NGO's to serve as civil governance institutions which could enable sustainability in agriculture. However with partnership between NGO's, corporations, government and official development agencies growth may be possible. It is possible to have diversified and resilient ecoagriculture systems by providing incentives for improved agricultural practices.

I. Make in India Efforts

If we want to handle growing food demand then investment has to be measurable with deep understanding and by giving incentives at all points in our agricultural systems. A much broader vision with bolder ideas are required that can nourish both people and the universe. All the world's leading scientists and thinkers need to come on the same platform for agricultural development to happen. Financially and politically agriculture has faced many crises. For good change to happen we must not allow the crisis to happen as it may be disastrous to agriculture. To create a more sustainable food and agriculture system we need to take more meaningful action. Many areas have now begun to shift. Therefore, if we focus on that and build on that than much can be done. There are many evidences that indicates the environmental awareness and utilization of waste is taking roots in industry and business. Countries like Australia, Japan, Republic of Korea and Singapore are the world leaders in managing wastes. They have developed new and innovative technologies for waste disposal by looking into environmental protection systems through proper institutions. Good progress has been made in waste management practices by other countries like People's Republic of China, India, Indonesia, Malaysia, Pakistan, Philippines and Thailand.

Multi-lateral and bi-lateral development agencies (including various United Nations agencies (UNDP, UNEP, UNIDO, WHO, ESCAP), World Bank, ADB and some donor countries) are offering both technical and final assistance for waste disposal in countries of the region. Innovative and "Make in India" efforts towards development are required for the indigenous capabilities in the countries of the region in terms of expertise, equipment manufacture, process technology guidelines, design, construction/installation, operation and maintenance of waste. It is vital that the developing countries take great care in controlling pollution and waste, in enhancing recycling and reuse and to minimize hazardous wastes. This needs to be done at the earliest. Due to extensive improvement in resource and environment management expensive clean-ups can be avoided in industrialized countries. These will help in reducing the ultimate costs and save resources. These reduce the levels of pollution and other negative impacts on the local environment. Small-scale raw material processing is often labour-intensive and widely dispersed but it needs high energy. Such dispersed industries could relieve big cities of some of their population and pollution pressures. They could provide non-farming jobs in the countryside, produce consumer goods that cater to local markets and can help in spreading environmentally sound technologies.

II. Promises and Risks of New Technologies

Technology brings a positive impact on the social, cultural and economic fronts of the nations and also the world community. They open up many opportunities which will raise the productivity and living standards of the rural people. It helps in improving the health and conserves the natural resources. However, at the same time it may also bring new hazards which may require an improved capacity for risk assessment and risk management.

Today, information technology is of much use. Processes based mainly on the advances made in micro-electronics and computer sciences have a lot of importance. With rapid means of communication it can help in improving the productivity, energy, resource efficiency and organizational structure of industry.

Ceramics and plastics contribute to energy and resource conservation, they require less energy to manufacture and are light in weight. For improving the environment there are many new technologies like biotechnology that may play a major role. Genetically improved products may improve human and animal health to a large extent. Scientists are working towards finding new drugs, new therapies and new ways to either prevent or control disease vectors. Energy derived from plants could be useful for non-renewable fossil fuels. There can be a major leap in agriculture by using newer high yielding crop varieties and those which are resistant to unfavourable weather conditions and pests. Recent researches have observed that biotechnology has been helpful in cleaner and more efficient alternatives to many wasteful processes and polluting products.

Using new techniques for treating solid and liquid wastes will definitely help in solving the problems of hazardous waste disposal. Recent developments in space technology will help in agriculture-based economies. Weather forecasting services provided through a satellite and communication network can help farmers in deciding when to plant, water, fertilize and harvest their crops. Remote sensing and satellite imagery would further help in the optimal use of the Earth's resources, permitting the monitoring and assessment of long-term trends in climatic change, marine pollution, rates in soil erosion and other aspects of plant cover. These new technologies and the Green Revolution will be helpful in interlinking agriculture, industry and services. This will help developments in all different sectors. In developed countries agriculture has become like an 'industry'. Among the major agriculture issues weather forecasting, storage and transport are taking the front seat. Tissue culture technique and genetic engineering has made possible to obtain plant strains which have the ability to fix nitrogen from the air, a development that would drastically affect the fertilizer industry but that would also reduce the threat of pollution by agrochemicals.

In agriculture practices the type of seed is very vital. The seed business has increased by chemical and energy industries that are providing new seeds which meet the specific local conditions and requirements. Specific fertilizers and pesticides are also necessary. All this needs good coordination between the various fields like research and development, production and marketing. This will help not to be dependent on a few crop varieties or on the products. However, sometimes these technologies may have negative impact on the environment. The large-scale production and widespread use of new materials may create health hazards (such as the use of gallium arsenate in the microchip industry). The green revolution has concerns if dependence is on a few crop strains and large doses of agrochemicals. New life forms produced by genetic engineering needs to be tested and an assessment should be done to check their impact on health, on the maintenance of genetic diversity and ecological balance. Only after the assessment is done they should be introduced into the market.

III. Strategies for Sustainable Industrial Development

Coordination between various resources with the environment into the industrial planning with government will be helpful. This will allow a steady reduction in the energy, resource content for future growth by increasing the efficiency of resource use, reducing waste and will help resource recovery and recycling.

a. Establish Environmental Goals, Regulations, Incentives and Standards

It is also necessary that industry, government and the public have clear benchmarks. The national governments have to take a lead in establishing a clear environmental goal and enforce environmental laws, regulations, incentives and standards for industrial enterprises. During the formation of policies it is important that priority must be given for better public health and to reduce industrial pollution and hazardous wastes. All this must bring an overall improvement for which a timely review and monitoring should be done.

The regulations should also take up issues related to transport and disposal of toxic substances so that air and water pollution are at its minimum. The laws need to be formed at the centre and followed by all state governments. Strict regulations are needed across national boundaries to avoid all sorts of pollution. This must bring responsibility to each country not to harm the health and environment of other nations.

b. Make More Effective Use of Economic Instruments

In industrial production due to inefficiency a large amount of waste is produced which causes pollution. To get rid of this pollution is very expensive and it is then that the industries make investments to improve their products, processes and to increase efficiency. This will then reduce the pollution. Whenever, there are economic incentives such investments increase. It is necessary to have regulations which can impose uniform performance standards to ensure that industry makes the investments necessary to reduce pollution and waste.

Many times much emphasis is not given to air and water and these resources are regarded as 'free'. However, it has been observed that these resources have caused enormous pollution and heavy costs to the industries. Due to this pollution, the community gets affected in the form of damage to human health, property and ecosystems. The industries may invest in measures to prevent the damages. If they have market for its product then they pass the costs to the consumer. Recently the industrialists are also replanting forests, restocking fish, rehabilitating land after mining. These might compensate the victims towards better health and improvement of property damage. However, ultimately the consumers may have to suffer in the form of tax or higher product cost.

It will be therefore, worthwhile if the enterprises invest in preventive, restorative and compensatory measures with various subsidies. To prevent the external damage many times subsidies are given to the industries especially in many developing countries. This in turn increases the investment. Again, it is the taxpayer who suffers by paying more. Sometimes even tax distortions take place which are not good.

In 1972, the member countries of OECD agreed to base their environmental policies on a Polluter Pays Principle (PPP). In an economic efficiency policy, PPP has encouraged the industries to internalize environmental costs and combine it with the product price. The state regulations in CMEA countries are run by the government which allows the environmental concerns to be taken into account. As environment becomes an integral part a thorough knowledge is needed for implementing various issues. Industries have many problems on international relations and environment. Sometimes there are problems due to trade barriers or difficulties in cooperation between authorities. It was seen that in the policies of OECD the guidelines on PPP discourage subsidies to the industries. Probably if the incentives can reduce pollution then it may help. Technological innovations and product redesigning can lead to safer products with efficient processes. Raw material recycling can also be promoted. All this could happen by using a more effective, integrated use of economic incentives and disincentives such as investment tax breaks, low-interest loans, depreciation allowances, pollution or waste charges and non-compliance fees.

The policy objectives should be made to increase the effectiveness of environment. The subsidies on raw materials, water supply and energy must promote the development of industry in rural areas. Governments should examine if existing economic policies, instruments or subsidies that are provided to various industry-based programmes and projects are contributing effectively for environmentally sound and resource efficient practices.

c. Broaden Environmental Assessments

All around the globe the countries feel that any major investment should go through an environmental impact assessment. The environmental assessment should apply to the products, projects, policies and programmes. These must include major policies involving macroeconomic finance as they have a significant impact on the environment. Countries in Asia and Latin America have adopted systems for environmental impact assessment. Monitoring needs to be done for quality checks. Setting up of an international assessment committee will help in the developing countries.

d. Encourage Action by Industry

Industry's response to pollution should be beyond compliance with certain regulations. This should have social responsibility and awareness of environment at all levels. All industrial enterprises, trade associations and labour unions should establish policies concerning resources and environmental management along with their legal compliance.

International trade associations are helpful in disseminating information. Associations like International Chamber of Commerce and the European Council of Chemical Manufacturers' Federation have taken lead roles in dealing with environmental issues. These should work towards sustainable development. The resources are required for environmental regulations by the small and medium-sized industries. These mostly include metal working, machine tools, printing, tanning and dying. Sometimes it becomes difficult for them due to the lack of resources. With newer innovations like micro-electronics it is possible to have inexpensive means in controlling the production process. Energy-saving biological systems are good to avoid pollution of different types.

There is also a need to have training programmes for the management and also workers. This will motivate to adopt clean technologies and plan strategies to keep environment clean. With cooperative efforts and joint research and development in the smaller firms on environmental issues it is possible to control pollution to a large level.

e. Increase Capacity to deal with Industrial Hazards

Some chemical products have improved health and life expectancies, increased agricultural production and comfort. All this has expanded the opportunities at the economic level. The industries and the chemical products can have very severe impact on the environment. However, problems of product and process pollution have come into the picture. A large amount of waste is generated which affects human health and the environment. With increase in chemical products and very complex large-scale technologies it has been observed that accidents have become very common. Some of the heavy metals and non-metallic minerals are highly dangerous to the health and the environment. It is to be seen now how to come out with innovative and safer technologies and systems. There are many chemicals in the environment that are known to cause cancer, birth defects and have long-term genetic effects which will take many decades to get decomposed.

e.i Chemicals

Out of the total world trade 10% trade is taken by chemicals. It is estimated that every year around 1,000-2,000 new chemicals enter the commercial market. But for this, it is vital that they undergo an adequate prior testing or evaluation. The database lacks data for nearly 80 per cent of the chemicals used in commercial products and processes inventoried under the Toxic Substances Control Act. With the new system of pre product testing by 1986, more than 500 chemicals and chemical products had been banned altogether or had their uses severely restricted in the country of origin. Many of the toxic and hazardous chemicals are withdrawn from clearance processes every year in the light of control agency concerns or are never submitted to national control agencies for clearance. In industrialized countries there are chemical control agencies which restrict or ban any toxic chemical. An appropriate review is done for any action to be taken. However, it has been observed that developing countries have not made such rule in their system while importing. All details should be provided to the importing country with the information so that they could decide to either ban or restrict the chemical if found harmful. Third World importers have no way to effectively control trade in chemicals that have been banned or severely restricted in exporting countries. Therefore, they need the infrastructure to assess the risks associated with chemical use. Looking into the danger of the chemicals it will be worth while that the commission must pass a legislation that all governments particularly those of the major chemical producing countries should be tested for health and environmental impact before being launched into the international markets. This needs to be done for each criteria and for the assessment procedures. Proper systems are also required for international sharing of the tasks and for the resources required. This will help in regulating the export of chemicals to developing countries for which authorization of domestic sale has not been sought or given. There is a need to increase consumer awareness. Governments should encourage the establishment of information centres on chemical products used by consumers and strengthen the international network of information exchange, assessment and data banks evolving in the UN and elsewhere.

It is very vital to force regulations on the packaging and labelling of the toxic chemicals. There must be clear directions and it should be put in specific common local language apart from English where the chemical has to be utilized. Lead must be taken by the consumer unions and other non-governmental organizations in collecting and distributing comparative risk information on the ingredients in consumer products such as cleaning agents and pesticides. Towards this there has been much development in the developing countries. A lot of responsibility lies on the trade union of individual country to influence the authorities and management so that such accidents do not happen and investment from companies should not be done where they do not follow acceptable standards. Technology development has improved environment in the industrial parts of the world. For business to happen in the international market there must be sharing of production technologies and resources. This may be politically difficult but not impossible if persued properly (Juul Bjerke, International Confederation of Free Trade Unions, WCED Public Hearing, Oslo, 24-25 June 1985).

It is the responsibility of the chemical producers and also the user industries to look into the source of the risks which are associated with the chemicals. They have to confirm that their products should meet the safety standards without any side effects on health and the environment. The rules for handling should be given on the labels.

e.ii. Hazardous Wastes

Industrialized countries generate about 90 per cent of the world's hazardous wastes. In 1984 around 325 million tons were generated worldwide. About 5 million tons of this was from the newly industrialized and developing areas of the world.

In OECD member countries alone, thousands of waste disposal sites exist, many of which require remedial action. Clean-up is very expensive. It is estimated that an amount of $10 billion for the Federal Republic of Germany, more than $1.5 billion for the Netherlands and $20-100 billion for the United States and $60 million for Denmark was utilized (1966; in dollars).

It will be worth while if government intervenes through regulatory action or financial support. Managing waste in developing countries suffers from a variety of problems. With heavy rains in the tropics the wastes get leached into the soils under the landfills. This causes them to overflow. In the absence of pretreatment of the wastes it could contaminate the water supply or cause local people to be directly exposed to the wastes. It is seen that generally the land-filling occurs close to industrial areas that are surrounded by poor neighbourhoods. Therefore, it is important to see that the objective of the policy must be to reduce the amount of waste generated and to transform an increasing amount into resources for use and reuse. Thus the volume of waste which needs to be treated or disposed through incineration, land disposal or dumping at sea will reduce. This is a major problem in the industrialized countries as well as developing countries where rapid industrialization is bringing the same severe problems of hazardous waste management.

There is danger as amount of wastes crossing national frontiers is increasing day by day. From 1962 to 1983 it was found that the waste transported in Western Europe for disposal to another country had doubled. This was around 250,000- 425,000 tons (1-2 per cent of the total hazardous wastes generated). In 1984, it was 4,000 shipments of hazardous wastes which were sent from the Netherlands to the German Democratic. International transport of wastes which was supposed to be disposed in the sea, either by incineration or dumping was approximately 1.8 million tons in 1983. Offshore dumping in the waters of the Pacific and the Caribbean has become common which is very dangerous. A keen survey needs to be done to dispose hazardous (including radioactive) wastes. Many international bodies are looking into this matter and problems might be solved by strengthening international cooperation in this area. As per OECD norms there should be strict controls on shipments to non member countries, prior notification should be given to the country and consent should be taken from the country of final destination. Extensive draft guidelines have been developed by UNEP, but there has been no effective mechanism to monitor or to control hazardous waste trade and dumping. It can be achieved with the support of various governments and international organizations to control the movement of hazardous wastes.

e. iii. Industrial Accidents

It is possible that accidents concerning toxic chemicals and radioactive materials occur in industries anywhere in the world. According to a survey carried out by the U.S. Environmental Protection Agency, 6,928 accidents of varying severity have occurred at U.S. plants between 1980 and 1965 (average of five a day).

There was a case in Mexico City in 1984 where liquid gas storage tanks had exploded and had killed 1,000 people and many thousands had become homeless. There was another tragedy in India at Bhopal where more than 2,000 people were killed and more than 200,000 were injured. Another incident where an accident took place at a plant in West Virginia in the United States resulted in emergency evacuation of residents because of many health problems. The accidental release in 1976 of the highly toxic and mutagenic chemical dioxin at Seveso, Italy, and the ensuing saga of drums of contaminated soil being passed around Europe, also showed that in industrial countries regulations can be evaded and minimum safety standards breached.

A fire at a warehouse of a chemical manufacturer in Basel, Switzerland (1986) had sent toxic fumes into France and the Federal Republic of Germany and released toxic chemicals into the river Rhine. This had adverse effect on the fish and many fish were killed. This affected the vital water supply in countries downstream, all the way to the Netherlands. After investigation the scientists have found this damage of riverine ecosystems will take a long time to come back to normal state. Incidents at Mexico City, Bhopal, Chernobyl and Basel in a short time have brought much public concern about industrial disasters. Looking into all these details it is suggested that there is a strong necessity to bring together the bilateral and regional cooperation. Even yearly thorough survey of industrial operations for hazardous chemicals will be worthy. Accordingly, regulations or guidelines can be prepared to enforce the safe operation of industrial plants.

Landfill gas is creating a major role in this field. Most landfill-gas-to-energy projects involve converting the harmful emissions into electricity and power homes. It is also used as a vehicle fuel and also as a substitute for mains in houschold gas supply. The waste hierarchy as laid out in European law states the ideal chain of events for wastes as reduce, reuse, recycle, energy recovery and dispose. Countries like Germany and the Netherlands have had efficient methods in place for years. In many places residents throw their household waste into one bin and leave it for the municipal authority to separate it. While streams of mixed MSW (municipal sewage waste) can be collected and then separated into the various components like recyclable items and organic waste. There are many separation schemes in effect across the world and it depends on each municipality as to what will work best. The collection of food scraps into a separate bin is important so that the organic waste does not end up in landfill. It also means that biowaste can be turned into compost or biodegraded in a safe manner without emitting harmful gases. Systems for separating glass bottles, aluminium cans and plastics also mean that recycling becomes easier, safer and more efficient. Bio-methanation is a solution for processing biodegradable waste. The way ahead is to set up smart cities. Civic bodies must redraw their vision in solid waste management and rework the strategies with recent changes in lifestyles. The garbage management in cities needs to be done again so that we can process waste and not landfill it. The quantity and characteristics of MSW in India and forecasting future waste generation is fundamental to successful waste management planning. The quantity of MSW generated depends on living standards, the extent and type of commercial activity, eating habits, population density, economic status, level of commercial activity, culture and city/region.

Management in India and new updated draft rules have recently been published. The MoEF issued MSW (Management and Handling) rules 2000 to ensure proper disposal of waste. Municipal authorities are responsible for implementing these rules and developing infrastructure for collection, storage, segregation, transportation, processing and disposal of MSW. Chandigarh is the first city to develop SWM in a planned way and has improved waste management as compared to other Indian cities.

Sorting Technology

People will always, either through ignorance or carelessness, throw their waste in the wrong bin every now and then. So, however good our separate collection schemes may be it will be still important to remember that it is not always practical to have them in place. It is vital to create a method that can take a mixed waste stream and then divide it into reusable, separate waste streams. One of the greatest innovations in waste technology is the invention of a sorter. Sorting technology comes in many forms, from water-based technology like ArrowBio which separates the organic fraction from recyclables, to the whirring, and whizzing sorting machines. When mixed waste is fed into a single stream recycling facility the process will include many processes which may include removal of larger items by hand separation of items by weight. This may mean items like metals, plastics, paper and glass to be sorted out from each other. Other methods include use of screens to separate items by size. Magnetic separation of metals like eddy current separators for aluminium ultraviolet optical scanners (Near Infra-red and Medium Infra-red) combined with targeted air jets that send items of certain types in separate collection bins e.g. PET and non PET plastics.

Many companies have done brilliant work in this field over the past few years. They are leading the way in the development and manufacture of these types of technologies. One example is France-based Pellenc ST which has recently launched a new MIR (Medium Infra-red) sorter which sorts paper according to its quality, and has improved its NIR (Near Infra-red) system to sort wood into category A and category B.

Bibliography

A.O. A.C. 1975.Association of agricultural chemists, official methods analysis, 12th Ed. Washington, D.C.

Abdullah N. and Gerhauser H. 2008. Bio-oil derived from empty fruit bunches. Fuel. 87: 2606–2613.

Abel A.M., Carnell A.J., Davis JA and Paylor M. 2003. The synthesis of buprenorphine intermediates by regioselective microbial N-and O-demethylation reactions using *Cunninghamella echinulata*. Enzyme Microb Technol. 33:743–748.

Agarwal R. 1998. 'India: the World's final dumpyard'. Basel Action News 1(l): 1998.

Aggarwal B. B. and Shishodia S. 2006. Molecular targets of dietary agents for prevention and therapy of cancer. Biochem Pharmacol. 71: 1397-421.

Agrawal, R. and Basappa, S.C. 1994. Fermentation of paddy malt mash to ethanol by mixed cultures of *Saccharomyces cerevisiae* and *Zymomonas mobilis* ZM4 with penicillin G. J Fermen Bioeng. 77 (2): 218-220.

Agrawal, R. and Basappa, S.C. 1996. Role of antimicrobial agents in simultaneous saccharification and fermentation of paddy malt mash to ethanol by mixed cultures of *Saccharomyces cerevisiae* PHO3 and *Zymomonas mobilis* ZM4. Biotechnol Lett. 18 (6): 673-678.

Agrawal, R. and Joseph, R. 2000 a. Bioconversion of alpha pinene to verbenone by resting cells of *Aspergillus niger*. Appl microbiol biotechnol. 3(53):335-337.

Agrawal, R. and Joseph, R. 2000 b. Optimization of conditions for the biotransformation of alpha pinene to verbenone by a *Penicillium* sp. JFST. 37(4): 430-432.

Agrawal, R., Deepika, N.U. and Joseph, R.1999.Strain improvement of *Aspergillus* sp. and *Penicillium* sp. by induced mutation for biotransformation of alpha-pinene to verbenol. Biotechnol Bioeng. 63 (2):249-52.

Agricultural Research Service.1956. Agriculture hand book no. 98 november. Chemistry and technology of citrus, citrus product; and byproducts. United States Department of Agriculture. Washington, D C.

Akanbi, W.B. and A.O. Togun 2002. The influence of maize stover compost and nitrogen fertilizer on growth, yield and nutrient uptake of Amaranth. Scientia Horticulture. 93:1-8.doi: 10.1016/S0304-4238(01)00305-3.

Akhtar M, Scott G M, Swaney RE and Kirk TK.1998. Overview of biomechanical and biochemical pulping research In: Eriksson, K-EL, Cavaco-Paulo A (Eds). Enzyme applications in fiber processing. Amer Chem Soc. Washington, DC. 15–27.

Al Dufour J, Bartram R B andVictor G.(Eds.). Book on Animal Waste, Water Quality and Human Health. January 2012.Publisher: IWA Publishing. ISBN: 9781780401232.ISBN: 9789241564519 (WHO).

Alam MZ, Muyibi SA, Mansor MF and Wahid R. 2007. Activated carbons derived from oil palm empty-fruit bunches: application to environmental problems. J Environ Sci. 19:103–108.

Alburquerque JA, Gonzálvez J, García D and Cegarra J. 2006a. Measuring detoxification and maturity in compost made from "alperujo", the solid by-product of extracting olive oil by the two-phase centrifugation system. Chemosphere. 64: 470–477.

Alburquerque JA, Gonzálvez J, García D and Cegarra J. 2006b. Effects of bulking agent on the composting of "alperujo", the solid by-product of the two-phase centrifugation method for olive oil extraction. Process Biochem. 41: 127–132.

Alibardi L and Cossu R. 2015. Composition variability of the organic fraction of municipal solid waste and effects on hydrogen and methane production potentials. Waste Management. 36: 147-155.

Allen SG, Schulman D, Lichwa J, Antal MJ, Laser M and Lynd LR. 2001. A comparison between hot liquid water and steam fractionation of corn fiber. Ind. Eng. Chem. Res. 40: 2934–2941.

Alvarenga P, Gonçalves AP, Fernandes RM, Varennes AD, Vallini G, Duarte E and Cunha-Queda AC.2008. Evaluation of composts and liming materials in the phytostabilization of a mine soil using perennial rye grass. Sci Total Environ. 406: 43–56.

American Chemistry Council. 2007. Association of post consumer plastic recyclers. United States national postconsumer plastics bottle recycling report.

Amo T, Atomi H and Imanaka T. 2002. Unique presence of a manganese catalase in a hyperthermophilic archaeon, *Pyrobaculum calidifontis* VA1. J Bacteriol. 184 (12): 3305-3312.

Anastasiadi M, Pratsinis H, Kietsas D, Skaltsounis AL and Haroutounian SA. 2010. Bioactive non-coloured polyphenols content of grapes, wines and vinification by-products: Evaluation of the antioxidant activities of their extracts. Food Res Tnt. 43: 805-813.

Andary C, Wylde R, Heitz A, Rascol JP, Roussel JL and Laffite C.1985. Poliumoside, a caffeic glycoside ester from Teucrium belion. Phytochem. 24: 362–364.

Andary C, Wylde R, Laffite C, Privat G and Winternitz F.1982.Structure of verbascoside and orobanchoside, caffeic acid sugar esters from Orobanche rapum-genistae. Phytochem. 21:1123-1127.

Andjelkoviæ M, Radovanoviæ B, Radovanoviæ A and Andjelkoviæ A. 2013. Changes in polyphenolic content and antioxidant activity of grapes Vranac during ripening. South Afr. J Enolgy and Vit. 34: 147- 155.

Ando H, Sakaki T, Kokusho T, Shibata M, Uemura Y and Hatate Y. 2000. Decomposition behavior of plant biomass in hot-compressed water. Ind Eng Chem Res. 39:3688–3693.

Andrade MA, Carlo P, Sean I, O'Donoghue CW, MuÈller and Peer Bork.2001.Comparison of ARM and heat Protein Repeats. J Mol Biol. 309: 1±18.

Angerosa F, Alessandro N d', Konstantinou P and Giacinto L Di.1995. GC-MS evaluation of phenolic compounds in virgin olive oil. J Agric Fd Chem. 43 (7):1802–1807.

Angerosa F, Nicola d'A, Panajotis K, Luciana Di G. 2012. Animal waste, water quality and human health. Ed. Al Dufour, Jamie Bartram, Robert Bos and Victor Gannon.Published on behalf of the World Health Organization by IWA Publishing, Alliance House, 12 Caxton Street, London SW1H, QS, UK.

Anjello R and Ranawana A.1996. Death in Slow Motion- India has become the dumping ground for the west's toxic waste'.Asiaweek.

Antonious GF. 2009. Pesticides, food contaminants and agricultural wastes J Environ Sci Health. Part A44. Part B. Appl Environ Microbiol. 2013.79 (12): 3590–3600.

Araceli Castaneda-Ovando, Hernandez Ma de LP, Hernandez Ma EP, Rodriguez J A and Galan-Vidal CA.2009. Chemical studies of anthocyanins: A review. Fd Chem.113:859–871.

Aravantinos Z, Constantina T, Vassiliki O, Christos DT.1994.Fermentation of orange processing wastes for citric acid production.J Sci Fd Agric. 65(1).117–120.

Arnold L K.1931. Agricultural wastes in industry. J Chem Educ.8 (12):2310.

Arvanitoyannis IS and Tserkezou P.2008a.Corn and rice waste: a comparative and critical presentation of methods and current and potential uses of treated waste. Int J Fd Sci Technol. 43: 958–988.

Arvanitoyannis IS and Tserkezou P. 2008b.Wheat, barley and oat waste: a comparative and critical presentation of methods and potential uses of treated waste. Int J Fd Sci Technol. 43: 694-725.

Arvanitoyannis IS and Varzakas TH. 2008c. Vegetable waste treatment: Comparison and critical presentation of methodologies. Crit Rev Fd Sci Nutr. 48:205–247.

Arvanitoyannis IS, Kassaveti A and Stefanatos S.2007.Current and potential uses of thermally treated olive oil waste. Int J Fd Sci Technol. 42:852–867.

Ashoush IS and Gadallah MGE. 2011. Utilization of mango peels and seed kernels powders as sources of phytochemicals in biscuit.World J Dairy Fd Sci. 6(1):35-42. ISSN 1817-308X© IDOSI Publications.

Atanasiu B. 2010. The role of bioenergy in the national renewable energy action plans: a first identification of issues and uncertainties. Intelligent energy Europe.1-21.

Atiq Uz Zaman. 2009. Life cycle environmental assessment of municipal solid waste to energy technologies. Global J Environ Res.3 (3): 155-163. ISSN 1990-925X, © IDOSI Publications.

Aviram M and Rosenblat M. 2012. Evidence-Based Complementary and Alternative Medicine. 2012:Article ID 382763, 20.pages.http://dx.doi.org/10.1155/2012/382763. ReviewArticle. Pomegranate Protection against Cardiovascular Diseases.

Ayyagari V, Mohapatra SC, Venkatramaiah A, Thiagasundaram T, Choudhuri D, Johri DC and Renganathan.1980. Selection for egg production on part records. Part1: evaluation of short term response to selection. Theor Appl Genet.57: 227-283.

Aziz NH, Farag SE, Mousa LAA and Abo-Ziad MA.1998. Comparative antibacterial and antifungal effects of some phenolic compounds.Microbios.93:43–54.

Bahukhandi D and Munjal RL. 1989. Cultivation of *Pleurotus* sp. on different agricultural residues. Indian Phytopath. 42 (4):492–495.

Balasundrama N, Sundram K and Samman S. 2006. Analytical, nutritional and clinical methods, Phenolic compounds in plants and agri-industrial by-products: Antioxidant activity, occurrence and potential uses. Fd Chem. 99:191–203.

Ball AS, Betts WB, McCarthy AJ.1989. Degradation of lignin-related compounds by actinomycetes. Appl Environ Microbiol. 55(6):1642–1644.

Bambidis VA and Robinson PH.2006.Citrus by-products as ruminant feeds: A Review.Animal Feed Sci Technol.128:175-217.

Banini AE, Boyd LC, Allen JC, Allen HG and Sauls DL.2006. Muscadine grape products intake, diet and blood constituents of nondiabetic and type 2 diabetic subjects. Nutri. 22:1137-1145.

Bano Z and Srivastava HC. 1962. Studies on cultivation of *Pleurotus* species on paddy straw.Fd Sci.12:363-365.

Barrero AF, Oltra JE, Raslan DS and Saude DA.1999. Microbial transformation of sesquiterpene lactones by the fungi *Cunninghamella echinulata* and *Rhizopus oryzae*. J Nat Prod. 62 (5):726–729.

Basappa SC, Somashekar D, Renu Agrawal, K Suma and K Bharathi. 1997. Nutritonal composition of fermented ragi (chang) by Phab and defined starter cultures as compared to unfermented ragi (*Eleucine caracanis* G). Intern J Fd Sci Nutri. 48: 313-319.

Baydar NG, Sagdic O, Ozkan G and Cetin S. 2006. Determination of antibacterial effects and total phenolic contents of grape (*Vitis vinifera*) seed extracts. Inter J Fd Sci. 41:799-804.

Beall DS, Ohta K and Ingram LO. 1991. Parametric studies of ethanol production from xylose and other sugars by recombinant *Escherichia coli*. Biotechnol Bioeng. 38: 296-303.

Beaudin N, Caron RF, Legros R, et al., 1999. Identification of the key factors affecting composting of a weathered hydrocarbon-contaminated soil. Biodegradation. 10:127-133.

Beg QK, Kapoor M, Mahajan L and Hoonda GS. 2001. Microbial xylanases and their industrial applications: A review. Appl Microbiol Biotechnol. 56:326-338.

Benavente-García O, Castillo J, Marin FR, Ortuño A and Del Río JA.1997.Uses and properties of Citrus flavonoids. J Agri Fd Chem. 45: 4505-4515.

Beri V, Sidhu B S, Bahl G S and Bhat A K.1995. Nitrogen and phosphorus transformations as affected by crop residue management practices and their influence on crop yield. Soil Use Manage. 11: 51-54.

Beri V, Sidhu B S, Bhat A K and Singh B P. 1992. Nutrient balance and soil properties as affected by management of crop residues. In: Proceedings of the international symposium on nutrient management for sustained productivity. Vol II (Ed: Bajwa M S) 133-135. Department of Soils, Punjab Agricultural University, Ludhiana, Punjab, India.

Bertelli AA, Giovannini L, Giannessi D et al. 1995.Antiplatelet activity of synthetic and natural resveratrol in red wine. Int J Tissue React.17:1–3.

Beyene G and Alemu A. 2015. Review of Mango (*Mangifera indica*) Seed-Kernel Waste as a Diet for Poultry. J Biol Agri Healthcare. 5 (11):156-159.

Bhandari TPS, Singh RN and Verma BL.1991. Cultivation of oyster mushroom on different substrates. Indian Phytopath. 44 (4):555–557.

Bhatia SK and Yang YH. 2017. Microbial production of volatile fatty acids: current status and future perspectives. Rev Environ Sci Biotechnol. 16:327–345.

Bhatti M, Mir FA and Siddiq M.1987. Effect of different bedding material on relative yield of oyster mushroom in successive flushes. Pak J Agric Res. 8:256–259.

Bierkland NK and Holo H.1993. Transduction of a plasmid carrying the cohesive end region from Lactococcus lactis bacteriophage LC3. Appl Environ Microbiol. 59:1966–1968.

Bignell E, Cairns T C, Throckmorton K, Nierman WC and Keller N P.2016.Secondary metabolite arsenal of an opportunistic pathogenic fungus. Biol Sci. DOI: 10.109.

Bilanovic D, Shelet G and Green M.1994. Xanthan fermentation of citrus waste. Bioresource Technol. 48: 169-172.

Bindman N A, Bobeica S C, Liu W R and van der Donk Wilfred A. 2015.Facile removal of leader peptides from lanthipeptides by incorporation of a hydroxy acid. J Am Chem Soc.137 (22):6975–6978.

Bio-Medical Waste (Management & Handling) Rules. 1998. Ministry Environment Forests. Notification. New Delhi.

Birdie Scott Padam, Tin H S, Chye F Y and Abdullah MI. 2014. Banana by-products: an under-utilized renewable food biomass with great potential. J Fd Sci Technol. 51 (12): 3527–3545.

Blanco P, Sieiro C and Villa TG.1999.Production of pectic enzymes in yeasts. FEMS Microbiol Lett. 175: 1-9.

Boateng Kofi K, George YO and Ebenezer M. 2017. Review Rice Cultivation and Green house Gas Emissions: A Review and Conceptual Framework with Reference to Ghana.Agriculture.7:7.doi:10.3390/agriculture.7010007 www.mdpi.com/journal/agriculture.

Bocco A, Cuvelier M E, Richard H and Berset C.1998. Antioxidant activity and phenolic composition of citrus peel and seed extracts. JAgri Fd Chem. 46: 2123-2129.

Bodie EA, Bower B, Berka RM and Dunn-Coleman NS. 994. Economically important organic acid and enzyme products. In: Martinella SD, Kinghoron JR (Eds) *Aspergillus* 50 Years of Progress in Industrial Microbiology, Elsevier, Amsterdam. The Netherlands. 562-568.

Bok S H, Jeong T S, Lee S-K, Kim J-R, Moon S S and Choi MS. 2001. Flavanone derivatives and composition for preventing or treating blood lipid level-related diseases comprising same. US Patent No. 20010006978A1.

Bolado-Rodríguez S, Toquero C, Martín-Juárez J, Travaini R and García-Encina P A. 2016. Effect of thermal, acid, alkaline and alkaline-peroxide pretreatments on the biochemical methane potential and kinetics of the anaerobic digestion of wheat straw and sugarcane bagasse.Biores Technol. 201:182-190.

Bolzonella D, Pavan P, Zanette M and Cecchi F. 2007. Two-Phase Anaerobic Digestion of Waste Activated Sludge: Effect of an Extreme Thermophilic Prefermentation. Ing. Eng. Chem. 46: 6650–6655.

Bonatti M, Karnopp P, Soares HM and Furlan SA. 2004. Evaluation of *Pleurotus ostreatus* and *Pleurotus sajorcaju* nutritional characteristics when cultivated in different lignocellulosic wastes. Food Chem. 88: 425 - 428.

Bouchard J, Nguyen TS, Chornet E and Overend RP.1991.Analytical methodology for biomass pretreatment. Part 2: characterization of the filtrates and cumulative product distribution as a function of treatment severity. Bioresour Technol. 36:121–131.

Bradamante, S., Barenghi, L. and Villa, A. 2004. Cardiovascular protective effects of resveratrol. Cardiovascular Drug Rev. 22: 169-188.

Bramorski A, Christen P, Ramirez M, Soccol CR and Revah S. 1998. Production of volatile compounds by the edible fungus *Rhizopus oryzae* during solid state cultivation on tropical agro-industrial substrates. Biotechnol Lett. 20(4):359–362.

Brar SK and Verma M. 2015. (Institut National de la Recherche Scientifique, Centre Eau, Terre & Environnement/Centre for Water, Earth and Environment, Québec, Canada and others.

Brennan TCR, Krömer JO and Nielsen LK. 2001. Physiological and Transcriptional Responses of Saccharomyces cerevisiae to d-Limonene show changes to the cell wall but not to the plasma membrane. Brinton WF. Biocycle. 42:74-76.

Buyuksonmez F, Rynk R, Hess TG and Bechinski E. 2000. Occurrence, degradation and fate of pesticides during composting. Part II: occurrence and fate of pesticides in compost and composting systems. Compost Sci Util. (8):61–81. doi: 10.1080/1065657X. 2000. 10701751.

Cáceres A, Pefia-Neira A, Galvez A, Obreque-Slier E, López-Solls R and Canals JM. 2012. Phenolic compositions of grapes and wines from cultivar Cabernet Sauvignon produced in Chile and their relationship to commercial value. JAgric Fd Chem.60: 8694-8702.

Cadahía E, Muñoz L, Fernández de Simón B and García-Vallejo MC. 2001. Changes in low molecular weight phenolic compounds in Spanish, French and American oak woods during natural seasoning and toasting. JAgri Fd Chem.49:1790–1798.

Capanema EA, Balakshin MY and Kadla JF. 2005.Quantitative characterization of a hardwood milled wood lignin by nuclear magnetic resonance spectroscopy. J Agric Fd Chem. 53: 9639–9649.

Capasso R, Evidente A, Schivo L, Orru' G, Marcialis MA and Cristinzio G.1995. Antibacterial polyphenols from olive oil mill waste waters. J Appl Bacteriol. 79:393-398.

Cardona CA and Sánchez O J. 2007. Fuel ethanol production: Process design trends and integration opportunities.Biores Technol. 98:2415-2457.

Carmona E, Moreno M, Avilés M, et al., 2012. Composting of wine industry wastes and their use as a substrate for growing soilless ornamental plants. Spanish J Agricul Res.10: 482-491.

Carrol KK, Kurowska EM and Guthrie N. 1999. Use of citrus limonoids and flavonoids as well as tocotrienols for the treatment of cancer. Intl. Patent WO9915167.

Castaño R, Borrero C and Avilés M. 2011. Organic matter fractions by SP-MAS 13C NMR and microbial communities involved in the suppression of *Fusarium* wilt in organic growth media. Biol Control. 58 (3):286–293.

Cavinato C, Da Ros C, Pavan P and Bolzonella D. 2017. Influence of temperature and hydraulic retention on the production of volatile fatty acids during anaerobic fermentation of cow manure and maize silage. Bioresour Technol. 223: 59–64.

Chai M, Qingshi Tu, Mingming Lu and Yang YJ. 2014. Esterification pretreatment of free fatty acid in biodiesel production, from laboratory to industry. Fuel Processing Technol.125: 106–113.

Chaiprasert P. 2011. Biogas Thailand. J Sustainable Energy Environ. Special Issue: 63-65.

Chakraborthy S, Bisht HC and Agrawal MD.1991. Studies on varietal screening of mangoes of Uttar Pradesh for their suitability for production of canned nectar, juice and pulp. Ind. Fd Packer. 45(5):49-57.

Chapla D, Divecha J, Madamwar D and Shah A.2010.Utilization of agro-industrial waste for xylanase production by *Aspergillus foetidus* MTCC 4898 under solid state fermentation and its application in saccharification. Biochem Eng J. 49(3):361–369.

Chau CF and Huang YL. 2003. Comparison of the chemical composition and physicochemical properties of different fibers prepared from the peel of *Citrus sinensis* L. cv. Liucheng. J Agri Fd Chem. 51: 2615-2618.

Chen J, Chen T and Zhang Y. 2012.Vanadium tailings for high performance ceramsite synthesis. Metal Mine. In Chinese. 427(1): 161-165.

Chen Shieh-Shung Tom and Chang CJ. 1998. Microbial transformation of taxol and cephalomannine. US 5756536.

Chen Y, Senesi N and Schnitzer M. 1977. Information provided on humic substances by E4 /E6 rations. Soil Sci Soc Amer J. 41: 352-358.

Chenoweth WL and Leveille GA.1975. Metabolism and physiological effects of pectins. In: Jeanes A, Hodge J (Eds). Physiological Effects of food carbohydrates. Amer Chem Soc. Washington DC.312-324.

Cherry JR and Findatsef AL. 2003. Directed evolution of industrial enzymes: an update. Curr Opin Biotechnol.14: 438-443.

Chidthaisong A, Inubushi K, Muramatsu Y, *et al.*, 1996. Microbes Environ.11: 73-78.

Christen P, Meza J and Revah S.1997. Fruity aroma production in solid state fermentation by *Ceratocystis fimbriata*: influence of the substrate type and the presence of precursors. Mycol Res.101(8):911–919.

Ciafardini G and Zullo BA.1998.Inhibitory activity of the oil-mill waste water on *Rhizobiummeliloti* and *Rhizobium hedysarii* in the soil. Adv Fd Sci. (CMTL). 20: 89-93.

Ciafardini G and Zullo BA. 2000. β-glucosidase activity in olive brine during the microbiological debittering process. Adv Fd Sci. (CMTL) 22: 69-76.

Ciafardini G and Zullo BA.2001. β-glucosidase activity in *Leuconostoc mesenteroides* associated with fermentation of "Coratina" cultivar olives. Italian J Fd Sci.13: 41-51.

Claasen PAM, van Lier JB, Lopez-Contreras AM, van Niel EWJ, Sijtsma L, Stams AJM, de Vries SS and Weusthuis RA.1999. Utilisation of biomass for the supply of energy carriers. Appl Microbiol Biotechnol. 52:741-755.

Claudia P, Vigani M and Rodríguez-Cerezo E. 2015.Agricultural Nanotechnologies: What are the current possibilities? Nano Today. 10(2):124-127.

Clemente R, Hartley W, Riby P, Dickinson NM and Lepp NW. 2010. Trace element mobility in a contaminated soil two years after field-amendment with green waste compost mulch. Environ Pollution. 158: 1644e1651.

Clemente R, Walker DJ and Bernal MP. 2005. Uptake of heavy metals by *Brassica juncea* grown in a contaminated soil in Aznalcóllar (Spain): the effect of soil amendments. Environ Pollution. 138: 46-58.

CNST Nanotechnology Workshop. 2013. 11th Annual CNST Nanotechnology Workshop at the University of Illinois.

Collins AM and Kennedy MJ. 1999. Biotransformations and bioconversions in New Zealand: past endeavours and future potential.Austral Biotechnol. 9(2):86-94.

Collins RJ and Miller RH.1979. Minerals and the Environment.1:8-19.Minerals Yearbook, Volume I: Metals, Minerals, and Fuels. Doi: 10.1007/BF02010592.

Conde E, Cadahia E,Garcia-Vallejo MC and Tomas- Barberan F. 1995. Low molecular weight polyphenols in wood and bark of *Eucalyptus globulus*.Wood Fiber Sci. 27: 379-383.

Conrad R. 2002. Control of Microbial Methane Production in Wetland Rice Fields.Nutrient Cycling in Agroecosystems. 64: 59-69.

Corrales M, Toepfl S, Butz P, Knorr D and Tauscher B. 2008. Extraction of anthocyanins from grape by-products assisted by ultrasonics, high hydrostatic pressure or pulsed electric fields: A comparison. Innovative Fd Sci EmergingTechnol. 9: 85–91.

Cotterchio M, Boucher BA , Manno M, Gallinger S, OkeyA and Harper P. Dietary Phytoestrogen Intake Is Associated with Reduced Colorectal Cancer Risk.J Nutr. 2006.136(12): 3046–3053.

Cozzolino D. 2015. Review Infrared Spectroscopy as a Versatile Analytical Tool for the Quantitative Determination of Antioxidants in Agricultural Products, Foods and Plants. Antioxidants. 4: 482-497.

Cruz JM. Dominguez JM, Dominguez H and Parajo JC. 2001. Antioxidant and antimicrobial effects of extracts from hydrolysates of lignocellulosic materials. JAgr Fd Chem. 49:2459-2464.

Current world environment, An International Research Journal of Environmental Science, ISSN: 0973-4929.

Dale K, Kirk M, Sinclair M, Hall R and Leder K. 2010. Reported waterborne outbreaks of gastrointestinal disease in Australia are predominantly associated with recreational exposure. Aust. N Z J. Public Health. 34: 527–530.

Danieli B, de Bellis P, Carrea G and Riva S. 1990. Enzyme-mediated regioselective acylations of flavonoid disaccharide monoglycerides. Helvetica Chimica Acta.73:1837-1844.

Das SK, Kumar S and Ramachandra Rao P. 2000. Exploitation of iron ore tailing for the development of ceramic tiles. Waste Management. 20 (8):725-729.

Davis L, Jeon YJ, Svenson C, Rogers P, Pearce J and Peiris P. 2005. Evaluation of wheat stillage for ethanol production by recombinant *Zymomonas mobilis*. Biomass Bioenergy. 29(1):49–59.

de Gregorio A, Mandalari G, Arena N, Nucita F, Tripodo MM and Lo Curto RB. 2002. SCP and crude pectinase production by slurry-state fermentation of lemon pulps.Biores Technol. 83:89-94.

de la Fuente C, Clemente R, Martínez-Alcalá I, Tortosa G and Bernal MP. 2011. Impact of fresh and composted solid olive husk and their water-soluble fractions on soil heavy metal fractionation, microbial biomass and plant uptake. J Hazardous Mat. 186: 1283 - 1289.

de Vries JW and Faubion JM.1999. Defining dietary fiber:A report on the AACC/ILSI NA consensus workshop. Cereal Fd World. 44:506-507.

Dekkers JC and Hospital F. 2002. The use of molecular genetics in the improvement of agricultural populations. Nat Rev Genet. 3:22–32.

Del Pozo-Insfran D, Brenes CH, Serna S and Talcott ST.2007.Polyphenolics and antioxidant capacity of white and blue corns processed into tortillas and chips. Cereal Chemis. 84 (2): 162–168.

Dellepiane D, Bosio B and Arato E. 2003. Clean energy from sugarcane waste: feasibility study of an innovative application of bagasse and barbojo. JPower Sources.122:47-56.

Demircia F, Yoshiaki N, Kýrýmera Nes¸'e and Baser K H Can. 2004. Microbial transformation of (-) carvone. Z. Naturforsch. 59: c, 389-392.

Deng Hongchao and Yachao Li.2011. Experiment of hot-pressing molding likeness flagstone with iron-tailings powder. J Jilin University (Earth Science Edition). In Chinese. 41 (1): 322-326.

Detroy RW and Hesseltine CW. 1977. Availability and utilisation of agricultural land agro-industrial wastes. 4483 Purchased by U.S. Dept. of Agriculture for Official Use. Presented at the American Institute of Chemical Engineers Symoosium. New York. New York.

deVries J W. 2003. Proceedings of the Nutrition Society. On defining dietary fibre. AACC, American Association of Cereal Chemists. CAB International PNS Proceedings of Nutrition Society (2003) 0029-6651© Nutrition Society 2003 621 PNS 234.General aspects of dietary fibre.62:37–43. DOI: 10.1079/PNS2002234.

Dhakal RP, Ghimire KN and Inoue K. 2005. Adsorptive separation of heavy metals from an aquatic environment using orange waste. Hydrometallurgy. 79:182-190.

Dhillon G and Kaur S. 2016. Agro-industrial wastes as feedstock for enzyme production , apply and exploit the emerging and valuable use options of waste biomassagro-industrial wastes as feedstock for enzyme production. 1st Edition, ISBN9780128023921. Editor(s): Dhillon and Kaur.

Dhillon SS, Rajwant KG, Gill SS and Singh M. 2004. Studies on the utilization of citrus peel for pectinase production using fungus *Aspergillus niger*.Inter J Environ Stud. 61 (2):199-210.

Diaz LF. 2011. Solid waste management in developing countries: status, perspectives and capacity building. Intergovernmental preparatory meeting for CSD-19. United Nations Headquarters New York, USA.

Dickinson E, Klinkhamer E, Thomas G and Winsten R. 2011. The Future of Fuel Cells in Australia.An Interactive Qualifying Project completed in partial fulfilment of the Bachelor of Science degree at Worcester Polytechnic Institute. Professor Holly Ault, Advisor Craig Memery, ATA Liaison Division 42. Project #HXA-A111.

Dikeman N.Cleantech. org. What is Cleantech? 2009. Cleantech.org Research Brief Energy Storage Venture Capital Investment and Valuation Analysis 2001-2009.

Dilta BS, Sharma BP, Baweja HS and Kashyap B.2011. Flower drying techniques.A review. Intern J Farm Sci.1 (2): 1-16.

Divyashree MS, Jijomon George and Renu Agrawal. 2006. Biotransformation of chiral terpenes by *A. niger* and *Pseudomonas putida*. J Fd Sci Technol. 43 (1): 73-76.

Doughari et al., 2008a. Antioxidant Activity of Lignin Phenolic Compounds Extracted from Kraft and Sulphite Black Liquors. ttps://www.researchgate.net/publication/49694554.

Doughari J H and Manzara S. 2008b. *In vitro* antibacterial activity of crude leaf extracts of *Mangifera indica* Linn. Afri J Microbiol Res. (2):067-072. http://www.academicjournals.org/ajmr ISSN 1996-0808 ©2008 Academic.

Draghia-Akli R, Malone PB, Hill LA, Ellis KM, Schwartz RJ and Nordstrom JL. 2002. Enhanced animal growth via ligand-regulated GHRH mygenic-injectable vectors. FASEB J. 16: 426–428.

Dsule PG, Sharma AK, Banerjee K and Karibasappa GS. 2012. Raisin industry in India: adoption of good drying practices for safe raisins. NRC for Grapes, Bulletin. 85: (974/975/976) 209–215.

Duah-Yentumi S and Kuwatsuka S. 1980. Effect of organic matter and chemical fertilizers on the degradation of benthiocarb and MCPA herbicides in soil. Soil Sci Plant Nutri. 26: 541-549.

Dukare A S, Prasanna R, Chandra Dubey S, Nain L, Chaudhary V, Singh R *et al.*, 2011. Evaluating novel microbe amended composts as biocontrol agents in tomato. Crop Prot. 30:436-442. 10.1016/j.cropro.2010.12.017.

Ehara K and Saka S. 2002. A comparative study on chemical conversion of cellulose between the batch-type and flow-type systems in supercritical water. Cellulose. 9:301–311.

Ehara K and Saka S. 2005. Decomposition behavior of cellulose in supercritical water, subcritical water and their combined treatments. J Wood Sci. 51: 148–153.

El-Fouly MZ, Sharaf AM, Shahin AAM, El-Bialy HA and Omara AMA. 2015. Biosynthesis of pyocyanin pigment by *Pseudomonas aeruginosa*. J Radiation Res Appl Sci. 8(1):36–48.

Erb R, Chew B and Keller H. 1977. Relative concentrations of estrogen and progesterone in milk and bold and excretion of estrogen in urine. J Animal Sci. 46: 617–626.

Eriksson KEL, Blanchette RA and Ander P.1990.Microbial and enzymatic degradation of wood and wood components. Springer-Verlag USA.407. ISBN: 354051600X. Publisher: Springer-Verlag, Berlin, Germany.

Ermawati R, Morimura S, Tang Y, Liu K and Kida K. 2007. J Biosci Bioeng. Degradation and behavior of natural steroid hormones in cow manure waste during biological treatments and ozone oxidation. 103(1):27-31.

European Commission — DG Environment — Waste in the EU. 08/09/2017

Ezejiofor TIN, Duru CI, Asagbra AE, Ezejiofor AN, Orisakwe OE, Afonne JO and Obi E.2012. Waste to wealth: production of oxytetracycline using streptomyces species from household kitchen wastes of agricultural produce. Afr J Biotechnol 11:10115–10124.

Faisal PA, Hareesh ES, Priji P, Unni KN, Sajith S, Sreedevi S, Josh MS and Benjamin S.2014. Optimization of parameters for the production of lipase from*Pseudomonas* sp. BUP6 by solid state fermentation. Adv Enzyme Res2:125–133.

FAO/IAEA (Food and Agriculture Organization of the United Nations/International Atomic Energy Agency). 2001. FAO/IAEA Mutant Varieties Database. http://www-infocris.iaea.org/MVD/.Accessed January 1, 2003.

Farrow R, Elkins E, Rose W *et al.*, 1969.Canning operations that reduce insecticide levels in prepared foods and in solid. Food Wastes. Residue Reviews 29:73.

Fei He, Liang NN, Lin Mu, Pan Q H, Wang J, Reeves M J and Duan CQ. 2012. Anthocyanins and their variation in red wines I. Monomeric anthocyanins and their color expression. Molecules. 17:1571-1601.

Fei Q, Michael T, Guarnieri LT, Lieve ML, Laurens N D and Philip TP. 2014. Bioconversion of natural gas to liquid fuel: Opportunities and challenges. Biotechnol Adv. 32(3):596-614.

Felsot AS and Dzantor EK.1990. Enhanced biodegradation of pesticides in the environment. Amer Chem Soc. 249-268.

Fernández-Bolaños José G, Óscar López M, Ángeles López-García and Azucena Marset. 2006. Chapter 20. Biological properties of hydroxytyrosol and its derivatives. www. intechopen. com

First NL.1991. New advances in reproductive biology of gametes and embryos. In: Petersen RA, McLaren A and First NL (Eds.). Animal applications of research in mammalian development. Current Commun Cell Mol Biol. Cold Spring Harbor, NY: Cold Spring Harbor Laboratory Press. 1–21.

Fletcher A (Ed). 2003. Gene identified to regulate milk content and yield. http:// foodproductiondaily. com/news/news-NG.asp?id=29318.

Floodgate G.1984. The fate of petroleum in marine ecosystems. In R.M. Atlas Petroleum Microbiology. Macmillion, New York, NY, USA.355-397.

Freking BA, Murphy SK, Wylie A A, Rhodes S J, Keele J W, Leymaster K A, Jirtle R L and Smith T P. 2002. Identification of the single base change causing the callipyge muscle hypertrophy phenotype, the only known example of polar overdominance in mammals. Genome Res. 12:1496–1506.

Frontera-Suau R, Bost FD, McDonald TJ *et al.*, 2002. Aerobic biodegradation of hopanes and other biomarkers by crude oil-degrading enrichment cultures. Environ Sci Technol. 36:4585-4592.Frutos I, Gárate A and Eymar E.2008. IV International Symposium on Ecologically Sound Fertilization Strategies for Field Vegetable Production. 261-268.

Frydman A, Weisshaus O, Huhman DV, Sumner LW, Bar-Peled M, Lewinsohn E *et al.*, 2005. Metabolic engineering of plant cells for biotransformation of hesperidin into neohesperidin, a substrate for production of the low-calorie sweetener and flavor enhancer. NHDC. J Agric Fd Chem. 53 (25):9708-9712.

Fu P, Gao Q, Wan W, Jiang R, Zhong C and Jiao L.1994. Chemical properties and anti-tumor activity of polysaccharides from roots of *Panax ginseng*. J Norman Bethune Uni Med Sci. 20:439–441.

Gaind S and Nain L. 2010. Corrigendum to "Exploration of composted cereal waste and poultry manure for soil restoration".Biores Technol.101:2996-3003.

Gaind S, Nain L and Patel VB. 2009. Quality evaluation of co-composted wheat straw, poultry droppings and oil seed cakes.Biodegradation. 20: 307-317.

Gajalakshmi S and Abbasi S A.2008. "Solid waste management by composting: State of the art", Crit Rev Environ Sci Technol.38 (5):311-400. DOI: abs/10.1080/10643380701413633.

Gales JR, Young TS, Willhite GP and Green DW. 1988. "Equilibrium Swelling and Syneresis Properties of Xanthan Gum-Cr (III) Gels," paper SPE/DOE 17328, presented at the SPE/DOE Enhanced Oil Recovery Symposium, Tulsa, Oklahoma.

Galloway SM, McNatty K P. *et al.*, 2000. Mutations in an oocyte-derived growth factor gene (BMP15) cause increased ovulation rate and infertility in a dosage sensitive manner. Nature Genetics. 25: 279–283.

Garay R M, Francisco Mac D, Acevedo M L, Calderón B and Araya J E. 2009. Crop residue use in particleboard. BioResources. 4(4):1396-1408.

Garima T and Singh S P. 2014. Application of Bioremediation on Solid Waste Management: A Review. J Bioremed Biodeg 5:248. doi:10.4172/2155-6199.1000248.

Garrido D, Rubin T, Poidevin M, Maroni B, Le Rouzic A, Parvy JP and Montagne. 2015. JPLoS Genet.18:11(2):1004995.

Garzon CG and Hours RA.1992.Citrus Waste: An alternative substrate for pectinase production in solid-state culture. Biores Technol. 39: 93-95.

Gassaraa F, Brara S K, Tyagia RD, Vermab M and Surampallic RY.2016. Screening of agro-industrial wastes to produce ligninolytic enzymes by *Phanerochaete chrysosporium*. Electr J Biotechnol.20:1–8.

Gaur AC.1999.Microbial Technology for composting of Agricultural residues by improved method. Ind council Agri Res. New Delhi. 43 - 53.

Gaur RS, Nishad J, Jakhar N and Kaur C. 2015. Food industry waste: mine of nutraceuticals.Intern J Sci Environ Technol. 4(1):205–229. 431.

Gayathri Balakrishnan and Renu Agrawal. 2014. Antioxidant activity and fatty acid profile of fermented milk prepared by *Pediococcus pentosaceus*. J Fd Sci Technol. 51(12): 4138–4142.

George B, Kaur C, Khurdiya DS and Kapoor HC. 2004. Antioxidants in tomato (*Lycopersiumesculentum*) as a function of genotype. Fd Chem. 84: 45–51.

Georges, M. 2001. Recent progress in livestock genomics and potential impact on breeding programs. Theriogenology. 55:15–21.

Ghosh, D. and Konishi, T. 2007. Anthocyanins and anthocyanin-rich extracts: role in diabetes and eye function.Asia Pac J ClinNutr. 16 (2):200-208.

Godden, B., Ball, A.S., Helvenstein, P. *et al.*, 1992. Towards elucidation of the lignin degradation pathway in actinomycetes. J Gen Microbiol. 138: 2441-2448.

Golueke, C.1991. The Art and Science of Composting. The Staff of BioCycle Journal of Waste Recycling. The JGPress Inc. Pennsylvania, USA.14-27.

Gomez FJ, Monasterio RP, Vargas VC, Silva MF. 2012. Analytical characterization of wine and its precursors by capillary electrophoresis. Electrophoresis. 33:2240–2252.

Gomez, F.J.V., Monasterio, R.P., Vargas, V.C.S. and Silva, M.F. 2012. Electrophoresis. Review. Analytical characterization of wine and its precursors by capillary electrophoresis. Electrophoresis. 33 (15): 2199–2474.

Gorinstein, M.B.O., Park YS, Haruenkit R, Lojek A, Ciz M, Caspi A, Libman I and Trakhtenberg S. 2001. Comparison of some biochemical characteristics of different citrus fruits. Fd Chem.74:309–315.

Goyal, S., Singh, D., Suneja, S *et al.*, 2009. Effect of rice straw compost on soil microbiological properties and yield of rice.Ind J Agricul Res. 43:263-268.

Grohmann, K. and Baldwin, EA.1992. Hydrolysis of orange peel with pectinase and cellulase enzymes. Biotechnol Lett. 14:1169-1174.

Grohmann, K. and Bothast R J.1994a. Pectin rich residues generated by processing of citrus fruits, apples and sugar beets. In: Enzymatic conversion of biomass for fuel production. (Eds.) Himmel ME, Baker JO and Overend RP.372-390. ACS Symp. Ser. No 566, Washington, DC: American Chemical Society.

Grohmann, K., Baldwin EA and Buslig BS.1994b. Fermentation of galacturonic acid and other sugars in orange peel hydrolysates by the ethanologenic strain of *Escherichia coli*. LON Ingram - Biotechnol Lett. 16: 281-286.

Grohmann, K., Cameron RG and Buslig BS.1995. Fermentation of sugars in organe peel hydrolysates to ethanol by recombinant *Escherichia coli* KO11. Appl Biochem Biotech.51/52: 423-435.

Grohmann, K., Manthey JA, Cameron RG and Buslig B ′ela S. 1998. Fermentation of galacturonic acid and pectin-rich materials to ethanol by genetically modified strains of *Erwinia*. Biotechnol Lett. 2(2):195-200.

Günther, K.B. 2016. New analytical approaches in natural product research and Department of pharmacognosy with medicinal plant unit of medical University of Lublin polish academy of sciences. 10th international symposium on chromatography of natural products the application of analytical methods for the development of natural products. Lublin (Poland) June 6-9.

Guo, H, Wenhua L, QingW, ChiL, YanHXia, XiangF and Xiaodong X.2007. Effect of Anthocyanin-Rich Extract from Black Rice (*Oryza sativa* L. *indica*) on Hyperlipidemia and Insulin Resistance in Fructose-Fed Rats. Plant Foods for Human Nutrition.62(1): 1–6.

Gupta, P.K. and Tsuchiya, T.1991.Chromosome Engineering in Plants: Genetics, Breeding Evolution. Amsterdam: Elsevier.

Gurpreet Dhillon and Surinder Kaur (Eds): 2016. Agro industrial wastes as feedstock for enzyme production (1st Edition) apply and exploit the emerging and valuable use options of waste biomass. EBook ISBN: 9780128026120.

Hale, EB. 1969. Domestication and the evolution of behavior. In: Hafez ESE(Ed).The Behavior of Domestic Animals. Baltimore: Williams and Wilkins 22–24.

Hameed, B.H., Tan IAW and Ahmad AL. 2009. Preparation of oil palm empty fruit bunch-based activated carbon for removal of 2,4, 6-trichlorophenol: optimization using response surface methodology. J Hazard Mater. 164: 1316–1324.

Hamelinck, C.N, Van Hooijdonk G and Faaij APC. 2005. Ethanol from lignocellulosic biomass: techno-economic performance in short, middle and long term. Biomass Bioenerg. 28: 384–410.

Han, DJ, Choi JS, Park M and Seo TS.1998. Biotransformation of (±)-Lavandulol by the Plant Pathogenic Fungus *Glomerella cingulate*. J Agric Fd Chem. 46 (9):3858–3862.

Hanselman TA, Graetz DA and WIlkie AC. 2003. Manure-borne estrogens as potential environmental contaminants: a review. Environ Sci Technol. 37:5471–5478.

Hansen M, Kristine AK, Sørensena BH and Björklund E.2011. Determination of ten steroid hormones in animal waste manure and agricultural soil using inverse and integrated clean-up pressurized liquid extraction and gas chromatography tandem mass spectrometry. Analytical Methods. (3): 1087-1095.

Harris G K, Gupta A, Nines R G, Kresty L A, Habib S G, Frankel W L, LaPerle K, Gallaher D D, Schwartz S J and Stoner G D. 2001. Effects of lyophilized blackraspberries on azoxymethane-induced colon cancer and 8-hydroxy-2'-deoxyguanosine levels in the Fischer 344 rat. Nutr Cancer. 40: 125-133.

Hattori R. 1981. Camphor tree and Ho wood essential oils. Koryo .134: 83-90.

Hautala K, Peuravuori J and Pihlaja K.2000. Measurement of aquatic humus content by spectroscopic analysis.Water Res. 34:246-258.

Hegazy ME F, Tarik AM, Abdelsamed IE, Abou-El-Hamd HM,Usama AM, Eman H R, Alaa M S, Wafaa AT, Abdelaaty AS, Khalid AS, Nahla SAA and Fayza MH.2015. Microbial biotransformation as a tool for drug development based on natural products from mevalonic acid pathway:A review. J Adv Res. 6(1): 17–33.

Hegazy ME, Mohamed TA, El Shamy AI, Mohamed AE, Mahalel UA, Reda EH, Shaheen AM, Tawfik WA, Shahat AA, Shams KA, Abdel-Azim NS and Hammouda FM.2015.Microbial biotransformation as a tool for drug development based on natural products from mevalonic acid pathway: A review. J Adv Res. 6(1): 17–33.

Heim, Tagliaferro and Bobilya.2002.Flavonoid antioxidants: chemistry, metabolism and structure-activity relationships. J Nutr Biochem.13 (10):572-584.

Herrara AA, Sanchez JM, Castro PA, Alvarez SH, Ferrer JEJ, Zamilpa A, Ramos RR, Monter HP and Tortoriello J. 2007. Clinical effects produced by a standardized herbal medicinal product of Hibiscus Sabdariffa on patients with hypertension, A randomized double blind,lisinopril controlled clinical trial.Planta Med.73:6-12.

Hertog MGL, Feskens EJM, Hollman PCH, Katan MB and Kromhout D. 1993. Dietary antioxidant flavonoids and risk of coronary heart disease.The zutphen elderly study. The Lancet.342:1007-1011.

Heyman Y, Vignon X, Chesne P, Le Bourhis D, Marchal J and Renard JP. 1998. Cloning in cattle: From embryo splitting to somatic nuclear transfer. Reprod Nutr Dev. 38:595–603.

Hirokazu N, Miyazawa M, Akazawa S and Kameoka H. 2015. Agricultural Nanotechnologies: What are the current possibilities? Nano Today.10 (2):124–127.

Hirokazu N, Miyazawa M, Akazawa S, and Kameoka H. 1998. Biotransformation of (±)-Lavandulol by the Plant Pathogenic Fungus *Glomerella cingulate*.J Agri Fd Chem. 46 (9): 3858-3862.

Hölker U, Höfer M, Lenz J. 2004. Biotechnological advances of laboratory scale solid state fermentation with fungi. Appl Microbiol Biotechnol.64:175–86.

Hotchkiss AT Jr, Estibaliz OM, William EG, Glenn RG and Robert AR. 2003. Pectic oligosaccharides as prebiotics. Oligosaccharides in food and agriculture.Chapter5:54–62.DOI:10.1021/bk-2003-0849.ch005.ACS Symposium Series,Vol.849. ISBN13:9780841238268eISBN: 9780841219533. American Chemical Society.

Hotta Y, Ezaki S, Atomi H and Imanaka T. 2002. Extremely stable and versatile carboxylesterase from a hyperthermophilic archaeon. Appl Environ Microbiol. 68 (8):3925-31.

Hue N and Liu J.1995. Predicting compost stability.Compost Sci Utilization. 3(2): 8-15.

Hue N, Ikawa H and Silva J. 1994. Increasing plant available phosphorus in an Ultisol with a yard waste compost.Commun Soil Sci PlantAnalysis. 25:3291-3303.

Hussein MZ, Tarmizi RSH, Zainal Z, Ibrahim R and Badri M. 1996. Preparation and characterization of active carbons from oil palm shells.Carbon. 34: 1447–1454.

Ionel, I. 2010. Clean technology from waste management.Advances in Waste Management.ISSN: 1790-5095. ISBN: 978-960-474-190-8.155-161.

Isci, A. and Demirer G. 2007. Biogas production potential from cotton wastes.Renew Energy. 32:750–757. doi: 10.1016/j.renene.2006.03.018

Ismail, A M S.1996.Utilization of orange peels for the production of multienzyme complexes by some fungal strains.Process Biochem. 31(7):645-650.

Izabela, K W. 2004. Anthocyanins- More than nature's colours. J Biomed Biotechnol. 5:239–240.

Jain ,A K and Vyas D. 2002. Yield response of *Pleurotus florida* on heat straw in combination with other substrate. Mush. Res.11: 19-20.

Jang, M, Cai L, Udeani GO, Slowing KV, Thomas CF, Beecher CW, *et al.*, 1997. Cancer chemopreventive activity of resveratrol, a natural product derived from grapes. Science. 275:218 – 220.

Janshekar H and Fiechter A.1983. Lignin: Biosynthesis, applications and biodegradation. In: Pentoses and lignin. Adv Biochem Eng/Biotechnol. 27:119–178.

Jayagopal V, Albertazzi P, Kilpatrick ES, Howarth EM, Jennings PE, Hepburn DA and Atkin SL. 2002. Beneficial effects of soy phytoestrogen intake in postmenopausal women with type 2 diabetes. Diabetes Care 25:1709–1714

Jayani RS, Saxena S and Gupta R. 2005. Microbial pectinolytic enzymes:A review. Process Biochem. 40: 2931-2944.

Jayathilakan K, Sultana K, Radhakrishna K and Bawa AS. 2012. Utilization of byproducts and waste materials from meat, poultry and fish processing industries: A Review. J Fd Sci Technol. 49: 278–293.

Jenkins J.1999. Compost Myths: To Turn or Not To Turn. The Humanure Handbook.A Guide to Composting Human Manure, The Humanure Handbook. Second Edition Printed in the United States of America. Published in the United States by: Joseph Jenkins, Inc., 143 Forest Lane Grove City, PA 16127 USA. josephjenkins.com.

Jeon SM,Kim HK, Kim HJ, Do GM, Jeong TS, Park YB *et al.,* 2007. Hypocholesterolemic and antioxidative effects of naringenin and its two metabolites in high cholesterol fed rats. Transl. Res. 149(1): 15-21.

Jogaiah S, Oulkar DP, Vijapure AN, Maske SR, Sharma AK and Somkuwar RG. 2013. Influences of canopy management practices on fruit composition of wine grape cultivars grown in semi-arid tropical region of India. Afr J Agri Res. 8 (26):3462- 3472.

Juul B.1985.International confederation of free trade unions.WCED Public Hearing, Oslo. 24-25 June.

Kang T H, Hur J Y, Kim H B, Ryu J H and Kim S Y. 2006. Neuroprotective effects of the cyanidin-3-O-beta-d-glucopyranoside isolated from mulberry fruit against cerebral ischemia. *Neurosci. Lett.* 391:122–126.

Kanotra S and Mathur RS.1994. Biodegradation of paddy straw with cellulolytic fungi and its application on wheat crop. Biores Technol. 47:185-188.

Karaca A, Turgay OC and Tamer N. 2006. Effects of a humic deposit (gyttja) on soil chemical and microbiological properties and heavy metal availability. Biol Fertil Soils. 42:585-592.

Karibasappa GS and Adsule PG. 2008. Evaluation of wine grape genotypes by National Research Center for Grapes at their Farm at Pune, Maharashtra (India). Acta Hort.785: 497-504.

Kaur G, Kumar S and Satyanarayana T. 2004. Production, characterization and application of a thermostable polygalacturonase of a thermophilic mould *Sporotrichum thermophils* Apinis. Biores Technol Barking. 94(3):239-243.

Kaur PS, Kaur S, Kaur H, Sharma A, Raj P and Panwar S. 2015. Solid substrate fermentation using agro industrial waste: new approach for amylase production by *Bacillus licheniformis*. Intern J Curr Microbiol Appl Sci. 4:712–717.

Kaur M, Singh R P, Gu M, Agarwal R and Agarwal, C. 2006. Grape seed extract inhibits *in vitro* and *in vivo* growth of human colorectal carcinoma cells. Clinical Cancer Res. 12: 194-202.

Khalid M, Ratnam CT, Chuah TG, Ali S and Choong TSY. 2008. Comparative study of polypropylene composites reinforced with oil palm empty fruit bunch fiber and oil palm derived cellulose. Mater Des. 29: 173–178.

Khan AS, Fiorotto ML, Hill LA, Malone PB, Cummings KK, Parghi D, Schwartz RJ, Smith RG and Draghia-Akli R. 2002. Maternal GHRH plasmid administration changes pituitary cell lineage and improves progeny growth of pigs.Endocrinology.143:3561–3567.

Kimber, RWL.1973.Phytotoxicity from plant residues. III. The relative effect of toxins and nitrogen immobilisation on the germination and growth of wheat. Plant soil. 38:543-555.

Kirk, T.K. 1983. Degradation and conversion of lignocelluloses. In: Smith J E, Berry DR and Kristiansen B. The filamentous fungi. Fungal Technol. London: Edward Arnold. 4: 266-295.

Kjeldsen, P.M. 2002. Present and Long term composition of MSW landfill leachate: A Review.Critical Rev Environ Sci Technol. 297-336.

Klayman DL.1985.Qinghaosu (artemisinin): an antimalarial drug from China. Science. 228:1049–1055.

Klein TM, Wolf ED, Wu R and Sanford JC.1987. High velocity microprojectiles for delivering nucleic acids into living cells.Nature. 327:70–73.

Klinkhamer J and Klean Industries. 2009. Forecast of the three main clean technology sectors.Knox A. 2005.An Overview of Incineration and EFW Technology as Applied to the Management of Municipal Solid Waste (MSW).

Kobayashi N, Okada N, Hirakawa A, Sato T, Kobayashi J, Hatano S, Itaya Y and Mori S. 2009. Characteristics of solid residues obtained from hot-compressed-water treatment of woody biomass. Ind Eng Chem Res. 48: 373–379.

Koeck R. *et al.*, 2014. Preparation and evaluation of monolithic poly (N-vinyl carbazole-co-1, 4-divinylbenzene) capillary columns for the separation of small molecules. Analytical Bioanalytical Chem. 406: 5897–5907.

Kondo JK and McKay LL.1984. Plasmid transformation of *Streptococcus lactis* protoplasts: Optimization and use in molecular cloning. Appl Environ Microbiol. 48:252–259.

Kozukue N, Misoo S, Yamada T, Kamijima O and Friedman M. 1999. Inheritance of morphological characters and glycoalkaloids in potatoes of somatic hybrids between dihaploid *Solanum acaule* and tetraploid *Solanum tuberosum*. J Agric Fd Chem. 47(10):4478-4483.

Kris-Etherton PM, Hecker KD, Bonanome A, Coval SM, Binkoski AE, Hilpert KF, *et al.*, 2002. Bioactive compounds in foods: their role in the prevention of cardiovascular disease and cancer.Am J Med.113:71S–88S.

Kruse A. 2009. Hydrothermal biomass gasification. J Supercrit Fluids. 47:391–399.

Kumar A and Duhan J S. 2011. Production and characterization of amylase enzymeisolated from *Aspergillus niger* MTCC-104 employing solid state fermentation.Intern J Pharam Biol Sci. 2: B250–B258.

Kumar A, Gaind S and Nain L. 2008. Evaluation of thermophilic fungal consortium for paddy straw composting. Biodegradation.19:395-402.

Kumar, P. *et al.*, 2009. Methods for Pretreatment of Lignocellulosic Biomass for Efficient Hydrolysis and Biofuel Production. Industrial and Engineering Chemistry Research. 48(8): 3713-3729.

Laine R M, Furgal J C, Doan P, Pan D, Popova V and Zhang X. 2016. Avoiding carbothermal reduction: Distillation of alkoxysilanes from biogenic, green and sustainable sources. Angewandte Chemie Internatl. 55 (3): 1065–1069.

Larios G J M and Garcia C Huitrón.1989. Endo-polygalacturonase production from untreated lemon peel by *Aspergillus* sp. CH-Y-1043. Biotechnol Lett.11 (10): 729–734.

Larkin P and Scowcroft W. 1981. Somaclonal variation: A novel source of variability from cell culture for plant improvement.Theor Appl Genet. 60:197–204.

Laser M, Schulman D, Allen SG, Lichwa J, Antal MJ and Lynd LR. 2002. A comparison of liquid hot water and steam pretreatments of sugar cane bagasse for bioconversion to ethanol. Biores Technol. 81: 33–44.

Lata, Shukla L and Surender S. 2008. Composting Agro residues with fungal inoculants ICAR News Res.5.

Lawson IYD, Muramatsu K and Nioh I. 1995. Effects of organic matter on the growth, nodulation and nitrogen fixation of soyabean grown under acid and saline conditions. Soil Sci Plant Nutr. 41(4):721-728.

Le Blanc D. 2015. Towards integration at last? The Sustainable Development Goals as a network of targets, Sustainable Development. DESA Working Paper No. 141.ST/ESA/ 2015/DWP/141. 1-17.

Lee SJ, Ahn JK, Kim SH, Kim JT, Han SJ, Jung MY and Chung IM. 2003. Variation in isoflavone of soybean cultivars with location and storage duration. J Agri Fd Chem. 51: 3382–3389.

Lee SJ, Umano K, Shibamoto T and Lee KG. 2005. Identification of volatile components in basil (*Ocimum basilicum* L.) and thyme leaves (*Thymus vulgaris* L.) and their antioxidant properties. Fd Chem.91:131–137.

Lee SM, Bong-Seok B, Park HW, Ahn N G , Cho BG , Cho YL and Kwak YS. 2015. Characterization of Korean Red Ginseng (*Panax ginseng* Meyer): History, preparation method and chemical composition. J Ginseng Res. 39(4):384–391.

Lemes AC, Sala L, Ores J C, Braga AR, Egea MB and Fernandes KF. 2016. A review of the latest advances in encrypted bioactive peptides from protein-rich waste. Int J Mol Sci. 16: 17(6). Pii: E950.

Leresche JE and Meyer HP. 2006. Chemocatalysis and biocatalysis (biotransformation): some thoughts of a chemist and a biotechnologist. Org Proc Res Dev.10:572–580.

Leticia X Lopez-M,Rosa M O Ros,Gerardo VA andHugo S G. 2009. Antioxidant activity, phenolic compounds and anthocyanins content of eighteen strains of Mexican maize.LWT-Fd Sci Technol. 42(6):1187-1192.

Leung A Y. 1977. In: Leung's encyclopedia of common natural in gredients used in food, drugs and cosmetics Third Edition. Published by John Wiley and Sons, Inc., Hoboken, New Jersey (2010).

Li L, Liu R, Ye M, Hu X, Wang Q and Bi K. 2006. Microbial metabolism of evodiamine by *Penicillium Janthinellum* and its application for metabolite identification in rat urine. Enzyme Microb Technol. 39:561–567.

Li Y, Changjiang G, Yang J, Wei J, Xu J and Cheng S. 2006. Evaluation of antioxidant properties of pomegranate peel extract in comparison with pomegranate pulp extract. Fd Chem. 96:254–260.

Lionel K and Arnold J.1931. Agricultural wastes in industry. Chem Educ. 8 (12):2310.

Liu JH, Chen YG, Yu BY and ChenY J. 2006. A novel ketone derivative of artemisinin biotransformed by *Streptomyces griseus* ATCC 13273. Bioorg Med Chem Lett.16:1909–1912.

Liu Sean X and Mamidipally PK. 2005. Quality comparison of rice bran oil extracted with d-limonene and hexane. 82 (2): 209-215.

Liujie H, Wilfred A and van der Hughes D H. 2016. Discovery and characterization of bicereucin, an unusual d-Amino acid containing mixed two component lantibiotic. Medical Institute and Roger Adams Laboratory, Department of Chemistry, University of Illinois at Urbana"Champaign, 600 S. Mathews Avenue, Urbana, Illinois 61801, United States.J Am Chem Soc. 138 (16):5254–5257.

Lua AC and Guo J. 2001. Preparation and characterization of activated carbons from oil-palm stones for gas-phase adsorption. Colloids Surf. A 179:151–162.

Luchansky JB, Muriana PM, Klaenhammer TR.1988.Application of electroporation for transfer of plasmid DNA to *Lactobacillus, Lactococcus, Leuconostoc, Listeria, Pediococcus, Bacillus, Staphylococcus, Enterococcus, and Propionibacterium*.Mol Microbiol. 2: 637–646.

Lucinda H and Cohen A IG. 2002. Small molecule analysis by MALDI mass spectrometry. Analyt Bioanalyt Chem. 373(7): 571–586.

Lukaszewski AJ. 2004. Chromosome manipulation and crop improvement. In: Encyclopedia Plant Crop Sci. New York: Marcel Dekker. 40 (1): 216-225.

LuqueRodrýguez JM, Luque deCastro MD and Perez-Juan P. 2007. Dynamic superheated liquid extraction of anthocyanins and other phenolics from red grape skins of wine making residues.Biores Technol. 98:2705–2713.

Madan M, Vasudevan P and Sharma S. 1987. Cultivation of *Pleurotus sajor*-caju on different wastes. Biol Wastes. 22: 241–250.

Madhumithah CG, Krithiga R, Sundaram S, Sasikumar CS, Guhathakurta S and Mammen CK.2011.Utilization of vegetable wastes for production of protease by solid state fermentation using *Aspergillus niger*.World J Agri Sci. 7: 550-555.

Mahalakshmi N and Jayalakshmi S.2016.Amylase, cellulase and xylanase production from a novel bacterial isolate *Achromobacter xylosoxidans* isolated from marine environment. Intern J Adv Res Biol Sci 3:230–233.

Maheshwari DK.2014. Composting for sustainable agriculture.Springer Int. Pub.Switzerland. ISSN 2352-474X; ISBN 978-3-319-08003-1.

Mahmoud AE and Ali MM. 2014. Fermented pomegranate (*Punica granatum*) peel extract as a novel anticancer agent targeting angiogenesis and metastasis. Int J Toxicol Pharmacol Res. 6(4): 60-66.

Mahomoodally MF and Ameenah GF. 2013. Fighting multidrug resistance with herbal extracts, essential oils and their components. Chapter 15 – Harnessing traditional knowledge to treat existing and emerging infectious diseases in Africa. 223–235.

Maiti S, Gallastegui G, Suresh G, Sarma S J, Brar S K, Drogui P, Le B Y, Buelna G, Verma M et S and Carlos R.2018. Hydrolytic pre-treatment methods for enhanced biobutanol production from agro-industrial wastes. Biores Technol. 249 : 673-683.

Maiti SK, Nandhini S and Das M. 2005. Accumulation of metals by naturally growing herbaceous and tree species in iron ore tailings. Intern J Environ Stud. 62(5): 595-603.

Majeskie JL. 1996. Status of United States dairy cattle.In: National Cooperative Dairy Herd Improvement Program, Fact Sheet K-7. Washington, DC: U.S. Department of Agriculture.

Majors RE. 2001. New chromatographic columns and accessories at the 2001. Pittsburgh conference, Part I. LC GC Eur. 14(5): 284-301.

Malys N, Wishart JA, Oliver SG and Mc Carthy JE. 2011. Protein production in *Saccharomyces cerevisiae* for systems biology studies.Methods Enzymol. 500:197-212. Doi: 10.1016/ B978-0-12-385118-5.00011-6.

Mamma D and Christakopoulos. 2008. Citrus peels: An excellent raw material for the bioconversion into value added products.Tree and forestry science and biotechnol. 2 (Spl1): 83-97.

Mamma D, Kourtoglou E and Christakopoulos P. 2008. Fungal multienzyme production on industrial by-products of the citrus processing industry. Biores Technol. 99(7): 2373-2383.

Manach C, MazurA and Scalbert A. 2005. Polyphenols and prevention of cardiovascular diseases.Curr Opin Lipidol.16: 77–84.

Mandalari G, Bennett RN, Bisignano G, Saija A, Dugo G, Faulds CB and Waldron KW. 2006. Characterization of flavonoids and pectin from bergamot (*Citrus bergamia* Risso) peel, a major byproduct of essential oil extraction. J Agric Fd Chem.54: 197-203.

Mandalari G, Bennett RN, Bisignano G, Trombetta D, Saija A , Faulds CB, Gasson MJ and Nar A. 2007. Antimicrobial activity of flavonoids extracted from bergamot (*Citrus bergamia* Risso) peel, a byproduct of the essential oil industry.J App Microbiol. 103(6): 2056–2064.

Maria U, Soto R, Brown K and Ross C F. 2012. Antioxidant activity and consumer acceptance of grape seed flour-containing food products.Intern J Fd Sci Technol. 47: 592–602.

Marín J, Moreno J and Hernández T et al., 2006. Bioremediation by composting of heavy oil refinery sludge in semiarid conditions. Biodegradation.17 (3): 251-261.

Martin Hansen, Kristine A. Krogh, Bent Halling-Sørensen and Erland Björklund. 2011. Analytical Methods: Determination of ten steroid hormones in animal waste manure and agricultural soil using inverse and integrated clean-up pressurized liquid extraction and gas chromatography-tandem mass spectrometry. 3:1087-1095.

Martin J G P, Porto E ,Corrêa C B , de Alencar S M, da Gloria E M , Cabral I S R , de Aquino L M. 2012. Antimicrobial potential and chemical composition of agro-industrial wastes. J Nat Prodt. 5:27-36.

McAnulty SR, McAnulty LS, Morrow JD et al. 2005.Effect of daily fruit ingestion on angiotensin converting enzyme activity, blood pressure, and oxidative stress in chronic smokers-Free radical. Taylor & Francis

McHughen A. 2000. Pandora's Picnic Basket: The potential and hazards of genetically modified foods. New York: Oxford University Press.

McKendry P. 2002. Energy production from biomass (part 2): conversion technologies. Bioresour Technol. 83: 47–54.

Medeiros ABP, Pandey A, Freitas RJS, Christen P and Soccol CR. 2000. Optimization of the production of aroma compounds by *Kluyveromyces marxianus* in solid-state fermentation using factorial design and response surface methodology. Biochem Eng J.6 (1):33–39.

Mehta S and Smith KR. 2002. Exposure atlas for household energy and indoor air pollution modeling component: predicting household pollution levels.World Bank Energy Sector Management Assistance Programme. Washington, DC.

Metre A and Kaushik N. 2013. Revisiting transesterification to produce biodiesel: Novel reactors, separation strategies and economic outlook. J Surface Sci Technol. 29(1-2): 15-33.

Meysami P and Baheri H. 2003. Pre-screening of fungi and bulking agents for contaminated soil bioremediation. Adv Environ Res.7: 881-887.

Miraje S Y, Amlepatil NM, Sahoo AK and Mote GV. 2015. Anthocyanin Extraction from Winery Waste Material: A Review.

Miraje SY, Amlepatil NM, Sahoo AK and Mote GV. 2015. Anthocyanin Extraction from Winery Waste Material: A Review: Mote. JIPBS. 2 (2): 218-221.

Mohamed-Elamir F. Hegazy, Mohamed TA, ElShamy AI, Mohamed A HH, Mahalel U A, Reda EH, Shaheen AM, Tawfik WA, Shahat AA, Khalid A S, Nahla SAA and Hammoud Fayza M. 2015. Microbial biotransformation as a tool for drug development based on natural products from mevalonic acid pathway: A review. J Adv Res. 6(1):17–33.

Mohimani H, Kersten R D, Liu WT, Wang M, Purvine SO, Wu Si *et al.*, 2014. Automated genome mining of ribosomal peptide natural products. ACS Chem. Biol.9 (7): 1545–1551.

MonaM. Rashad, Abeer E. Mahmoud,, Mamdouh M. Ali, Mohammed U. Nooman and Amr S.Al-Kashef. 2015. Antioxidant and anticancer agents produced from Pineapple waste by solid state fermentation. Intern J Toxicol Pharmacol Res.7 (6): 287-296.

Mondal AK, Sengupta S, Bhowal J and Bhattacharya DK. 2012. Utilization of fruit wastes in producing single cell protein.Intern J Sci Environ Technol. 1:430 – 438.

Munch E W. 2007. Degumming of plants oils for different applications. 2005. Decision Support for Waste k. Minimization in wine-making processes.Wiley Inter Science DOI: 10.1002 353.

Musee N, Lorenzen L and Aldrich C. 2005. A fuzzy-based heuristic methodology for waste minimization analysis: A case study in the wine industry, Proceedings in the 7th World Congress of Chemical Engineering, Glasgow, Scotland, ISBN 0-852-9549- 4.

NAAB (National Association of Animal Breeders). 2000. Breakthroughs in biotechnology: Research equips producers with an array of genetic improvement tools. Online, available at http://www.naab-css.org/education/biotech.html.

Nankai H, Miyazawa M, Akazawa S and Kameoka H. 1998.Biotransformation of (±)-Lavandulol by the Plant Pathogenic Fungus Glomerella cingulate. J. Agric. Food Chem. 46 (9): 3858–3862

Nascimento TP, Sales AE, Camila CS, Romero RMP, Takaki GMC, Teixeira JAC, Porto TS and Porto ALF. 2015. Production and characterization of new fbrinolytic protease from *Mucor subtillissimus* UCP 1262 in solid-state fermentation. Adv Enzyme Res 3:81–91.

Nazhat Ul Ainn and Renu Agrawal.2003.Studies on solubilization and stabilization of alpha pinene hydroxylase from *Aspergillus niger*. J Fd sci Technol. 40 (1): 63-66.

Ni M, Leung DYC, Leung MKH and Sumathy K.2006.An overview of hydrogen production from biomass. Fuel Process Technol. 87:461–472.

Notter DR.1999. The importance of genetic diversity in livestock populations of the future.J Anim Sci.77:61–69.

NRC (National Research Council). 2002. Animal Biotechnology: Science-Based Concerns. Washington, DC: The National Academies Press.

Nunez-Selles. 2005. Antioxidant therapy: myth or reality? J Braz Chem Soc. 16 (4): 699-710.

Nychas GJE, Tassou CC and Board RG.1990. Phenolic extract from olives inhibition of *Staphylococcus aureus*. Lett Appl Microbiol. 10: 217-220.

Ok YS, Usman ARA, Lee SS, Abd El-Azeem SAM, Choi B, Hashimoto Y and Yang JE. 2011. Effects of rapeseed residue on lead and cadmium availability and uptake by rice plants in heavy metal contaminated paddy soil. Chemosphere. 85:677-682.

Olano-Martin E, Mountzouris K C, Gibson G R and Rastall R A. 2000. *In vitro* fermentability of dextran, oligodextran and maltodextrin by human gut bacteria. Br J Nutr. 83:247–255.

Pan I, Bomba D and Sen S K. 2012. Composting of common organic wastes using microbial inoculants. 3 Biotech. 2(2): 127–134.

Pandey A, Biswas S, Sukumaran RK and Kaushik N. 2009a. Study on Availability of Indian biomass resources for exploitation: a report based on a nationwide survey. TIFAC, New Delhi.

Pandey A, Gaind S, Ali A, *et al.*, 2009b. Effect of bioaugmentation and nitrogen supplementation on composting of paddy straw. Biodegrad. 20:293-306.

Pandey A, Soccol CR, Nigam P, Soccol VT, Vandenberghe LPS and Mohan R. 2000. Biotechnological potential of agro-industrial residues. II: cassava bagasse.Bioresour Technol. 74(1):81–87.

Panel Juliet Ben-Iwo, VasilijeManovic and PhilipLonghurst. 2016. Biomass resources and biofuels potential for the production of transportation fuels in Nigeria. Renewable and Sustainable Energy Reviews. 63:172-192.

Pant D, Reddy UG and Adholeya A. 2006. Cultivation of oyster mushrooms on wheat straw and bagasse substrate amended with distillery effluent.World J Microbiol Biotechnol. 22 (3): 267–275.

Papoutsopoulou SV, Ekateriniadou LV and Kyriakidis DA.1994. Genetic Construction of *Xanthomonas Campestris* and Xanthan Gum Production from Whey. Biotechnol Lett. 16(12):1235-1240.

Paré T, Dinel H and Schnitzer M. 1999. Extractability of trace metals during co-composting of biosolids and municipal solid wastes. Biol Fertil Soils. 29 (1):31-37.

Paredes C, Bernal MP, Cegarra J and Roig A. 2005. Influence of olive mill waste water in composting and impact of the compost on a Swiss chard crop and soil properties. Environ Internl. 31: 305-312.

Paredes C, Bernal MP, Roig A, Cegarra J, Sánchez-Monedero MA. 1996. Influence of the bulking agent on the degradation of olive-mill wastewater sludge during composting. International Biodeterioration and Biodegradation. 38 (3-4):205-210.

Parisi C, Vigani M and Rodríguez-Cerezo E.2014. Proceedings of a workshop on "Nanotechnology for the agricultural sector: from research to the field", Joint Research Centre.

Parisi C, Vigani M and Rodríguez-Cerezo E. 2015. Agricultural Nanotechnologies: What are the current possibilities? Nanotoday. Science Direct. 10 (2): 124–127.

Parr S and Bolwell.2000.Phenols in the plant and in man. The potential for possible nutritional enhancement of the diet by modifying the phenols content or profile. J Sci Fd Agric.80: 985-1012.

Paszczynski A and Crawford RL.1995. Potential for bioremediation of xenobiotic compounds by white rot fungi rot fungus *Phanerochaete chrysoporium*. Biotechnol Prog. 11:368–379.

Patnaik RS, Louie S, Gavrilovic V, Perry K, Stemmer WPC, Ryan CM and del Cardayre S. 2002. Genome shuffling of *Lactobacillus* for improved acid tolerance. Nat Biotechnol. 20:707–712.

Patrabansh S and Madan M.1997.Studies on cultivation, biological efficiency and chemical analysis of *Pleurotus sajor*-caju (Fr.) Singer on different bio-wastes.Acta Biotechnol.17:107-122.

Patricia G S, Aranzazu M S, Segura-Carretero A and Fernández-Gutiérrez A. 2010. Phenolic compound extraction systems for fruit and vegetable samples. Molecules.15:8813-8826.

Paul E A, Collins H P and Leavitt S W. 2001. Dynamics of resistant soil carbon of Midwestern agricultural soils measured by naturally occurring C-14 abundance. Geoderma.104: 239–256.

Payne K, Rico-Munoz E and Davidson P M.1989. The antimicrobial activity of phenolic compounds against *Listeria monocytogenes* and their effectiveness in a model milk system. J Fd Prot.52: 151-153.

Perez-Murcia MD, Moreno-Caselles J, Moral R, Perez-Espinosa A, Parades C and Rufete B. 2005. Use of composted sewage sludge as horticultural growth media: Effects on germination and trace element extraction. Commun Soil Sci Plant Anal. 36: 571–582.

Perner H, Schwarz D, Bruns C, Mäder P and George E.2007. Effect of arbuscular mycorrhizal colonization and two levels of compost supply on nutrient uptake and flowering of pelargonium plants. Mycorrhiza 17: 469–474 10.1007/s00572-007-0116-7.

Pfafflin JR and ý Ziegler E N. 2006. Encyclopedia of Environmental Science and Engineering. Fifth Edition.(1-2): CRC Press Book.

Philippoussis AN. 2009. Production of mushrooms using agro-industrial residues as substrates. Biotechnology for agro-industrial residues utilisation. Springer, Netherlands.163–196.

Pina A, Calderon IL and Benitex T. 1986. Intergeneric hybrids of *Saccharomyces cerevisiae* and *Zygosaccharomyces* fermentation obtained by protoplast fusion. Appl Environ Microbiol. 51:995–1003.

Planning Commission, Government of India. 2014 Report of the task Force on waste to energy (I) in the context of integrated municipal solid waste management.

Ponni C and Arumugan S. 2007. Effect on certain organic manures and Bio-stimulants on the growth and yield of *Phyllanthus niruri*. In: The Asian J Hort. 2(2):148-150.

PonnuswamyV, Vincenta P, Samuel G, Valan A, Mariadhas y Al-d and Abdullah N. 2016. Bioconversion of agro-industrial wastes for the production of fibrinolytic enzyme from *Bacillus halodurans* IND18: Purification and biochemical characterization. Electron J Biotechnol. 19 (2):1-8.ISSN0717-3458.

Prabha TN, Renu Agrawal, Ramasarma PR and Patwardhan MV. 1991. Intermediary metabolism and thymol formation in *in vitro* cultures of *Carum copticum*.Biotechnol Appl Biochem.14: 256-264.

Premsuda S, Achara C, Somporn M and Suparp A. 2012. Prebiotic isomaltooligosaccharide production from economic crops of Thailand.KKU. Res J.17:794-799.

Prince SPM, Sivakumar WS, Ravi V and Subburam V. 2000. The effects of coir pith compost on the growth and quality of leaves of the mulberry plant *Morus alba* L. Biores Technol. 72: 95–97.

Pushpa Khanna, OP Sharma, M Sehgal, C Bhargava, M Jain, A Goswami, S Singhvi, U Gupta, Renu Agrawal, P Sharma and SC Jain .1980. Antimicrobial principles from tissue culture of some plant species.Indian J Pharma Sci. 42: 113-120.

Puupponen-Pimia et al., 2001. Antimicrobial properties of phenolic compounds from berries.J Appl Microbiol. 90(4):494-507.

Qaim M and Kouser S. 2013. Genetically Modified Crops and Food Security.PLoSOne.8 (6):e64879.

Qiao L, Zhou YZH, Qi XL, Lin LH, Chen HCH and Pang LY. 2007. Biotransformation of cinobufagin by *Cunninghamella elegans*. J Antibiot. 60 (4):261–264.

Ragazzi E, Veronese G and Guiotto A.1973. Demethyloleuropein, a new glycoside isolated from ripe olives.Ann Chim. 63:13-20.

Rahman KHA, Yusof SJHM and Zakaria Z. 2016. Bioproteins production from palm oil agro-industrial wastes by *Aspergillus terreus* UniMAP AA-1 Pertanika. J Trop Agric Sci 39:29–39.

Raman D R, Williams E L, Layton A C, Burns RT, Easter JP, Daugherty AS, Mullen MD and Sayler GS.2004. Estrogen content of dairy and swine wastes.Environ Sci Technol. 38(13):3567-3573.

Rana R, Ganguly R, Gupta AK.2015.An assessment of solid waste management system in Chandigarh city India. Electron. J. Geotech. Eng. 20:1547–1572

Rao DN and Mikkelsen DS. 1977. Effect of rice straw additions on production of organic acids in a flooded soil. Plant and Soil.47 (2):303-311.

Rao N, Grethlein HE and Reddy CA.1996. Effect of temperature on composting of atrazine amended lignocellulosic substrates.Compost Sci Utilization. 4: 83–88.

Rashad MM, Mahmoud A E, Ali MM, Nooman M U, Al-Kashef AS. 2015. Antioxidant and Anticancer Agents Produced from Pineapple Waste by Solid State Fermentation.Intern J Toxicol Pharmacol Res. 7(6):287-296.

Rashad MM, Mahmoud AE and Ali MM.2011. Hepatoprotective, anti-inflammatory and analgesic potential of methanolic extract of fermented okara in experimental animals. Intern J Integra Biol. 11(3): 117-126.

Rathbone DA, Lister DL and Bruce NC.2001. Biotransformation of alkaloids.Alkaloids Chem Biol. 57:1-74.

Rati Rao E, S.V.N Vijayendra, M.C Varadaraj, Nirmala Devi, Renu Agrawal, Krishnanand and M.S Prasad. 2003. Fermentation technologies for smaller communities. J Rural Technol. 1 (1).

Ravani A and Joshi DC. 2013. Trends in post harvest technology.Mango and it's by product utilization–a review. 1 (1):55-67.

Renu Agrawal and M.V Patwardhan. 1987. Thymol A Secondary product from callus cultures of Ajowan (*Carum copticum*). J Fd Sci Technol. 24: 322-324.

Renu Agrawal and M.V Patwardhan/ 1993. Production of peroxidase enzyme by callus cultures of *citrus aurantifolia*.J Sci Fd Agri. 61: 377-378.

Renu Agrawal and Patwardhan M V. 1994. Formation of piperine from *in vitro* culture of *Piper nigrum* (Paniyoor) L. Ind J Plant Physiol. 37 (3): 171-173.

Renu Agrawal and Richard Joseph. 2000a. Factors affecting bioconversion of alpha pinene to verbenone by resting cells of *Aspergillus niger*.Appl Microbiol Biotechnol. 53: 335-337.

Renu Agrawal and Richard Joseph. 2000b. Optimization of conditions for the biotransformation of alpha pinene to verbenone by a soil isolate *Penilillium* sp. J Fd Sci Technol.37: (4). 430-432.

Renu Agrawal and Shylaja Dharmesh. 2011. An antishigella protein from *Pediococcus pentosaceous* MTCC 5151.Turkish journal of Biology. 36: 177-185.

Renu Agrawal, B.D Mohanty, N.S Singh and M.V Patwardhan.1990. Parameters affecting allicin formation as secondary metabolites by static cultures of *Allium sativum*. J Sci Fd Agricul. 57: 155-162.

Renu Agrawal, Deepika and S Vimala. 2000d. Conversion of nerol to dihydrocarvyl acetate by *Mucor* sp. Ind J Microbiol. 40: 49.

Renu Agrawal, Nazhat Ul A, Deepika and Richard Joseph. 1999. Improvement in the biotransformation of alpha pinene to verbenol in *Aspergillus* sp. and *Penicillium* sp. induced by mutation using chemicals and ultra violet irradiation.Biotechnol Bioengi. 63: 249-252.

Renu Agrawal, Patwardhan MV and Gurudutt KN. 1991. Formation of flavouring constituents in callus cultures of lime, *Citrus aurantifolia* S. Biotechnol Appl Biochem. 14: 265.

Renu Agrawal, Rati Rao ER, Vijayendra SVN, Varadaraj MC, Prasad MS and Krishna Nand. 2000. Flavour profile of idli batter prepared for defined microbial starter cultures.World J Microbiol Biotechnol.16: 687-690.

Renu Agrawal, Richard Joseph and Gurudutt KN.1996.Biotransformation of alpha -pinene to flavour volatiles by *Aspergillus niger*.Indian J Microbiol. 36: 121-122.

Renu Agrawal, Seetharam YN, Chandrakanth R, Kelamani and Jyothishwaran G. 2003. Biotransformation of ferulic acid to vanillin by locally isolated bacterial cultures. Ind J Microbiol. 2: 610-612.

Renu Singh, Vishal Kapoor and Vijay Kumar. 2010. Utilization of Agro-industrial Wastes for the simultaneous production of amylase and xylanase by thermophilic actinomycetes.Biochem Eng J. 49(3):388–394.

Rexroad CE. 1994. Transgenic farm animals. ILAR J. 36:5–9.

Rinker, D.L. and Alm, G. 2002. Evaluation of spawn and its role in green mold disease. Mushroom News. 50(7): 18–19.

Rivas B, Torrado A, Torre P, Converti A and Domínguez JM. 2008. Submerged citric acid fermentation on orange peel autohydrolysate.JAgric Fd Chem.56:2380–2387.

Rodriguez A, Serrano L, Moral A, Perez A and Jimenez L. 2008. Use of high-boiling point organic solvents for pulping oil palm empty fruit bunches. Bioresour. Technol. 99: 1743–1749.

Romero C, García P, Brenes M, García A and Garrido A. 2002. Phenolic compounds in natural black Spanish olive varieties. Eur. Fd Res. Technol.215: 489–496.

Rossi S, Vandenberghe L, Pereira B, Gago F, Rizzolo J, Pandey A, Soccol C and Medeiros A. 2009. Improving fruity aroma production by fungi in SSF using citric pulp. Fd Res Int 42(4):484–486.

Rozman HD, Saad MJ and Ishak ZAM. 2003. Flexural and impact properties of oil palm empty fruit bunch (EFB)–polypropylene composites–the effect of maleic anhydride chemical modification of EFB. Polym. Test .22: 335–341.

Rozman HD, Tay GS, N M Kumar, Mohd. Ishak ZA. 2001a. T1-Polypropylene-oil palm empty fruit bunch-glass fibre hybrid composites: A preliminary study on the flexural and tensile properties. European Polymer Journal. 37:128- 1291.

Rozman, HD, Tay GS, Abubakar A and Kumar RN. 2001b. Tensile properties of oil palm empty fruit bunch–polyurethane composites. Eur Polym J. 37: 1759–1765.

Ruanglek V, Maneewatthana D and Tripetchkul S. 2006. Evaluation of Thai agro-industrial wastes for bio-ethanol production by *Zymomonas mobilis*.Process Biochem. 41(6): 1432–1437.

Rudra SG, Nishad J, Jakhar N and Kaur C. 2015.Food industry waste: Mine of nutraceuticals. Internat JSci EnvironTechnol.4(1):205–229. ISSN 2278-3687.

Ruhul Amin M, Saquib Hossain M and Sarker M. 2013. Simulation of Ethanol Production by Fermentation of Molasses. J Eng (JOE) 1(4):69-73.ISSN: 2325-0224.

Ryan-Borchers Tracy A, Park Jean Soon, Chew Boon P, McGuire Michelle K, Fournier Lisa R and Beerman Kathy A. 2006. Soy isoflavones modulate immune function in healthy postmenopausal women. Am J Clin Nutr. 83:1118 –25.

Sadh P K, Duhan S and Duhan J S. 2018. Agro-industrial wastes and theirutilization using solid state fermentation: a review. Biores Bioprocess. 5:1-15.

Saini S, Singh J and Saini M. 2014. An improved branch and bound algorithm for cyclopeptide sequencing.Intern J Bioinform Comput Biol. 2 (1): 15-19.

Salciccioli M, Weiting Y, Mark AB, Jingguang G C and Dionisios GV. 2011. Differentiation of O–H and C–H bond scission mechanisms of ethylene glycol on Pt and Ni/Pt using theory and isotopic labeling experiments. JAm Chem Soc.133 (20):7996–8004.

Salimi M, Safari F, Tavasoli A and Alireza Shakeri. 2016. Hydrothermal gasification of different agricultural wastes in supercritical water media for hydrogen production: a comparative study.Internat J Indust Chem.7(3):277–285.

Samman S, Lyons Wall PM and Cook NC.1998. Flavonoids and coronary heart disease (Dietary perspectives). In: C.A. Rice-Evans, L. Packer (Eds.) Flavonoids in Health and Disease. Marcel Dekker, New York, USA.469–482.

Sánchez et al., 2002. Mushroom Biology and Mushroom Products. (Eds). UAEM. ISBN 968-878-105-3.

Sanford LL, Kowalski SP, Ronning CM and Deahl KL.1998. Leptines and other glycoalkaloids in tetraploid *Solanum tuberosum* × *Solanum chacoense* F2 hybrid and backcross families. Am J Potato Res. 75:167–172.

Saritha M, Arora A, Singh S *et al.*, 2012. Pretreatment of paddy straw with *Trametes hirsute* for improved enzymatic saccharification. Biores Technol. 1011-1040.

Sarma SJ, Dhillon G, Hegde K and Brar S K. 2012. Utilization of agro-industrial waste for the production of aroma compounds and fragrances.Chapter 10. In book: Biotransformation of agro-industrial wastes: Fine biochemicals, Publisher: Springer.

Sarmah A K, Meyer M T and Boxall A B A. 2006. A global perspective on the use, sales, exposure pathways,occurrence, fate and effects of veterinary antibiotics (VAs) in the environment. Chemosphere.65 725–759.

Sasaki K, Noparatnaraporn N, Nagai S and Martin A.1991. Use of photosynthetic bacteria for the production of SCP and chemicals from agroindustrial wastes. Bioconversion of waste materials to industrial products. Elsevier, New York, NY. 225–264.

Sasaki M, Adschiri T and Arai K. 2003. Fractionation of sugarcane bagasse by hydrothermal treatment. Bioresour Technol. 86: 301–304.

Satish Y M, Nikhilesh M A, Sahoo A K and Mote GV. 2015. Anthocyanin Extraction from Winery Waste Material: A Review.JIPBS. 2 (2): 218-221.

Sayara T, Borràs E, Caminal G et al., 2011. Internat Biodeterioration Biodegrad. 65: 859-865.

Schiber A, Berardini N and Carle R. 2003. Identification of flavonol and xanthol glycosides from mango peels by HPLC.JAgri Fd Chem.51:5006-5011.

Scott WE, McKay HH, Schaffer P, Fontaine TD. 1949. The partial purification and properties of antibiotic substances from the banana (*Musa sapientum*).JClin Investigation.28(5 Pt 1),899. http://dx.doi.org/10.1172/JCI102175.

Sears ER. 1981. Transfer of alien genetic material to wheat. In: Evans LT, Peacock WJ. (eds.) Wheat Science, Today and Tomorrow. Cambridge: Cambridge University Press.75–89.

Sears ER. 1956. The transfer of leaf rust from Ae. Umbellulata to wheat. Brookhaven Symp Biol. 9:1–22.

Sebastian SA and Chaleff RS. 1987. Soybean mutants with increased tolerance for sulfonyl urea herbicides.Crop Sci 27:948–952.

Seidel GE Jr. 1984. Applications of embryo transfer and related technologies to cattle. J Dairy Sci. 67:2786–2796.

Seyis I and Aksoz N. 2005. Xylanase Production from *Trichoderma harzianum*. FdTechnol Biotechnol. 43(1): 37–40.

Shah S A A, Tan H L, Sultan S, Faridz M A B M, Shah M A B M, Nurfazilah S and Hussain M.2014. Microbial catalyzed biotransformation of multifunctional triterpenoids derived from phytonutrients.Int J Mol Sci. 15(7):12027–12060.

Sharholy M, Ahmad K, Mahmood G, Trivedi RC. 2005. Analysis of municipal solid waste management systems in Delhi – a review. In Book of Proc. for the second Int. Congress of Chemistry and Environment, Indore, India, pp. 773–777.

Sharma J, Upadhyay A K, Adsule PG, Ramteke S D. 2012. Effect of climate change on grape and its value-added products.In book: Climate-Resilient Horticulture: Adaptation and Mitigation Strategies, Publisher: Springer India. (Eds.): H P Singh, N K Srinivas Rao and K S Shivashankar.67-80.

Shieh-Shung Tom Chen, Ching-jer Chang. 1998. US 5756536.Type: Grant Filed: October 17, 1996 Date of Patent: May 26, Assignee: Purdue Research Foundation.

Shin MJ, Kim YS, Kwak YS, Song YB, Kim YS and Park JD. 2004. Enhancement of antitumor effects of paclitaxel (taxol) in combination with red ginseng acidic polysaccharide (RGAP).Planta Med.70:1033–1038.

Shin WH, Park SJ and Kim EJ. 2006. Protective effect of anthocyanins in middle cerebral artery occlusion and reperfusion model of cerebral ischemia in rats. Life Sci.79:130–137.

Shipp J and El-Sayed MAA. 2010. Food applications and physiological effects of anthocyanins as functional food ingredients. The Open Fd Sci J. 4:7-22.

Shobharani P and Renu Agrawal. 2007. Therapeutic importance of volatile compounds produced by *Leuconostoc paramesenteroides*. Turkish J Biol. 31:35-40.

Shobharani P and Renu Agrawal. 2008. Effect on cellular membrane fatty acids in the stressed cells of *Leuconostoc mesenteroides* sub sp. Dextranicum. A native probiotic lactic acid bacterium. Fd Biotechnol. 22: 47-63.

Shobharani P and Renu Agrawal.2010 a.Interception of quorum sensing signal molecule by furanone to enhance shelf life of fermented milk.Fd Control. 21 (1): 61-69.

Shobharani P and Renu Agrawal. 2010 b. Enhancement of cell stability and viability of probiotic *Leuconostoc mesenteroides* MTCC 5209 on freeze drying to be used in food formulation.Internat J Fd Sci Nutri. 60 (56): 70-83.

Shobharani P, Ramesh B and Renu Agrawal. 2006. Strain improvement by mutagenesis and optimum conditions for culture parameters by response surface methodology for lactose tolerance in a novel native culture isolate *Leuconostoc mesenteroides* sub sp. Res J Biotechnol. 1 (2): 5-11.

Shrikhande AJ.2000.Wine by-products with health benefits. Fd Res Intern. 33:469-474.

Shymal Rajak S.C. Mahapatra and Basu M. 2011.Yield, fruit body diameter and cropping duration of oyster mushroom (*Pleurotus sajor* caju) grown on different grasses and paddy straw as a substrate. Eur J Med plants. 1 (1):10-17.

Sidhu B S and Beri V. 1989. Effect of crop residue management on the yields of different crops and on soil properties. Biol Wastes. 27(1): 15-27.

Sidhu B, Rupela O, Joshi PK and Beri V. 1998. Sustainability implications of burning rice-and wheat-straw in Punjab. Economic and Political Weekly. 33(39): 163-168.

Silva D, Martins ES, Silva R and Gomes E. 2002. Pectinase production from *Penicillium viridicatum* RFC3 by solid state fermentation using agricultural residues and agro-industrial byproducts.Braz J Microbiol. 33:318-324.

Sing NJ, Sing TC and Devi MB. 2003. Nutritional composition, processing and preservation of the edible mushroom found in Manipur using certain agro-waste and forest by product for Sustainable development. J Mycopathl Res.41 (2):221-223.

Singh R, Kapoor V and Kumar V. 2012. Utilization of agro-industrial wastes for the simultaneous production of amylase and xylanase by thermophilic actinomycetes.Braz J Microbiol. 43(4):1545–1552.

Singh R, Kapoor V and Kumar V. 2010. Utilization of agro-industrial wastes for the simultaneous production of amylase and xylanase by thermophilic Actinomycetes. Biochem Eng J. 49(3):388–394.

Singh S and Nain L. 2014. Microorganisms in the conversion of agricultural wastes to compost. Proc Indian Natn Sci Acad.80 (2): Spl. Sec. 473-481.

Singh S, Singh B, Mishra B et al., 2012. Microorganisms in sustainable agriculture and biotechnology. Springer.Netherlands.127-151.

Singhal S and Pandey S. 2001. Solid waste management in India: status and future directions. TERI Inf. Monitor Environ Sci. 6:1–4.

Sirkeci AA, Gul A and Bulut G. 2006. Recovery of Co, Ni and Cu from the tailings of divrigi iron ore concentrator.Mineral Process Extractive Metallurgy Rev. 27(2):131-141.

Sluiter A, Hames B, Ruiz R, Scarlata C, Sluiter J, Templeton D and Crocker D. 2005. Determination of structural carbohydrates and lignin in biomass.The US national renewable energy laboratory technical report.

Smith DC, Beharee V and Hughes JC. 2001. The effects of composts produced by a simple composting procedure on the yields of Swiss chard (*Beta vulgaris* L.) and common bean (*Phaseolus vulgaris* L.) Scientia Horticulturae 91:393-406.

Smitha Rao CV, Rati Rao E and Renu Agrawal. 2003. Enhanced production of verbenol, a high valued food flavourant by an intergeneric fusant strain of *Aspergillus niger* and *Penicillium digitatum*.Biotechnol Appl Biochem. 37 (2): 145-147.

Smolinska U J. 2000. Survival of *Sclerotium cepivorum* Sclerotia and *Fusarium oxysporum* Chlamydospores in soil amended with cruciferous residues. J Phytopath.148: 343 – 349.

Soares M, Christen P, Pandey A and Soccol CR. 2000. Fruity flavour production by *Ceratocystis fimbriata* grown on coffee husk in solid-state fermentation. Process Biochem. 35(8): 857–861.

Someya S, Yoshiki Y and Okubo K. 2002. Antioxidant compounds from bananas (*Musa cavendish*). J Fd Chem. 79: 351–354.

Somshekar D, Venkateshwaran G, Renu Agrawal, Prakash MH and Basappa SC.1999. Novel enrichment technique for the isolation of highly potent catalase producing yeast from soil. Biotechnol Techni.13: 65-68.

Soong K and Sutasinee S. 2013. Mango seed kernel oil and its physicochemical properties.Intern Fd Res J. 20(3):1145-1149.

Soundar S and Chandra TS. 1987. Production of cellulase and detection of Avicel-adsorbing carboxymethylcellulase from a mesophilic fungus. J Indl Microbiol 2:257-265.

Srokol Z, Bouche AG, van Estrik A, Strik RCJ, Maschmeyer T and Peters JA. 2004. Hydrothermal upgrading of biomass to biofuel: studies on some monosaccharide model compounds. Carbohydr. Res. 339: 1717–1726.

Stecher G. et al., 2011. LC-MS as a method of identification and quantification of plant metabolites. High Performance Liquid Chromatography in Phytochemical Analysis. 257-285.

Steenson LR and Klaenhammer TR.1987. Conjugal transfer of plasmid DNA between Streptococci immobilized in calcium alginate gel beads. Appl Environ Microbiol. 53: 898–900.

Strazzullo G, Schiano Moriello V, Poli A, Immirzi B, Amazio P and Nicolaus B. 2003. Solid wastes of tomato-processing industry (*Lycopersicon esculentum* "Hybride Rome") as renewable sources of polysaccharides. J Fd Technol.1 (3):102-105. ISSN 1993-6036.

Stredansky M and Conti E.1999. Xanthan production by solid state fermentation.Process Biochem. 34 (6–7): 581-587.

Strom PF. 1985. Effect of temperature on bacterial species diversity in thermophilic solid-waste composting.Appl Environ Microbiol.50: 899–905.

Strom PF. 1998. Evaluating pesticide residues in yard trimmings compost. Biocycle 39(11): 80.

Sudha N, Shobharani P and Renu Agrawal. 2006. Studies on the stability and viability of a local probiotic isolate *Pediococcus pentosaceous* (MTCC 5151) under induced gastrointestinal tract conditions.J Fd Sci Technol. 43 (6): 677-678.

Sudhakar T, Karpagam S and Premkumar J. 2015. Biosynthesis, antibacterial activity of pyocyanin pigment produced by *Pseudomonas aeruginosa* SU1.J Chem Pharma Res. 7(3):921-924.

Sueli Oliveira, Sandra Maria Gomes da Casta and Edmar element. 2002. Chemical composition of *Pleurotus pulmonarious* (Fr.) Quel, substrates and residue after cultivation. Brazilian Archives Biol. Tech.45:531-535.

Suganthi R, Benazir JF, Santhi R, Kumar RV, Hari A, Meenakshi N, Nidhiya KA, Kavitha G and Lakshmi R.2011. Amylase production by *Aspergillus niger* under solid state fermentation using agro-industrial wastes. Intern JEng Sci Technol 3:1756–1763.

Sukesan S and Watwood ME.1998. Effects of hydrocarbon enrichment on trichloroethylene biodegradation and microbial populations in finished compost. J Appl Microbiol. 85: 635–642.

Sulistyo J, Handayani R and Sudaryati SY. 2016. Biosynthesis and antioxidant activity of single cell carotenoids extracted from microbial cultures.JIPBS. 3 (1):133-142.

Suman B C and Sharma VP. 2005. Mushroom-cultivation and uses.Agribiose. Pub. House, Jodhapur.

Sun J, Chu YF, Wu X and Liu RH. 2002. Antioxidant and antiproliferative activities of fruits. JAgric Fd Chem.50: 7449-7454.

Sundar, S. *et al.*, 2014. Electron beam pretreatment of switchgrass to enhance enzymatic hydrolysis to produce sugars for biofuels. Carbohydrate Polymers. 100: 195-201.

Sung S H, Kyoung H K, Byong T J, Sun H C, Park J H, Kim D H, Kweon H J and Moon S H.2012.Antibacterial and antioxidant activities of tannins extracted from agricultural by-products. JMedPlants Res. 6(15): 3072-3079.

Survase DM. 2012. Bioconversion of agro waste into edible protein rich mushroom by *Pleurotus sajor* caju (fr.) singer. Trends Biotechnol Res. Vol. 1 No. 1 :60-62.ISSN 2320–0421(Print); ISSN 2320–043X(Online).

Surya Chandra Rao, Rachappa S B and Renu Agrawal. 2007. Role of oxygen scavengers in improving the stability and viability of probiotic lactic acid bacteria. Res J Biotechnol. 2 (1): 26-32.

Swanson EB, Couman MP, Brown GL, Patel JD and Beversdorf WD. 1988. The characterization of herbicide tolerant plants in *Brassica napus* L after *in vitro* selection of microspores and protoplasts. Plant Cell Rep. 2:83–87.

Tanaka R, Kosugi A and Mori Y. 2007. Utilization of lignocellulosic biomass discharged from palm oil industries. J Jpn Inst Energy. 86: 368–370.

Tanzi CD, Vian MA, Ginies C, Elmaataoui M and Chemat F. 2012. Terpenes as green solvents for extraction of oil from microalgae.Molecules.17:8196-8205;doi:10.3390/molecules17078196.ISSN1420-3049.

Tartoura K A H and Youssef S A. 2011. Stimulation of ROS-scavenging systems in squash (*Cucurbita pepo* L. var. Hybrid fadwa) plants by compost supplementation under normal and low temperature conditions. Scientia Horticulturae. 130: 862–868.

Tassou CC and Nychas GJE. 1995. Inhibition of Salmonella enteritidis by oleuropein in broth and in a model food system. Lett Appl Microbiol. 20: 120-124.

Tchobaanoglous G, Theisen H and Samuel A Vigil. 1993. Solid Waste Management in Urban Areas of Ghana: Issues and Experiences from Wa. Integrated Solid Waste Management, Open access journal. J Environ Pollution Human Health. 2014, 2 (5): 110-117.

Teixeira A, Baenas N, Dominguez-Perles R, Barros A, Rosa E, Moreno DA and Garcia-Viguera C. 2014. Natural Bioactive Compounds from Winery By-Products as Health Promoters: Int J Mol Sci. A Review. 15(9): 15638–15678.

Termorshuizen AJ, van Rijn E, van der Gaag DJ *et al.*, 2007. Suppressiveness of 18 composts against 7 pathosystems: variability in pathogen response. Soil Biol Biochem. 38: 2461–2477.

Thakur SA, Nemade SN and Sharanappa A. 2015. Solid state fermentation of overheated soybean meal (waste) for production of protease using *Aspergillus oryzae*. Inten J Innov Res Sci Eng Technol. 4:18456–18461.

Thampi SC, Wani PV, Bachchhav MB, et al., 1996. Nat. Symp. on Integrated disease manag. Rahuri, India: Indian Soc Pl Pathol. Studies on different *Pleurotus* species: Spawn production and morphological features. 53–54.

Thomsen M H. 2005. Complex media from processing of agricultural crops for microbial fermentation.Appl Microbiol Biotechnol. 68:598–606.

Thomsen M, Lassen P, Dobel S, Hansen PE, Carlsen L and Morgensen BB. 2002. Characterization of humic materials of different origin: A multivariate approach for quantifying the latent properties of dissolved organic matter. Chemosphere. 49:1327-1337.

Tiquia SM and Tam NFY. 2000. Microbiological and chemical parameters for evaluating the maturity of spent pig-litter compost.Warman P.R. (Ed). Cba Press (2000). Paper from Proc. Int. Composting Symp. (ICs '99). (1 and 2): 648-669.

Tommonaro G, Segura RCS, Santillana M, Immirzi B, De Prisco R, Nicolaus B and Poli A. 2007. Chemical composition and biotechnological properties of a polysaccharide from the peels and antioxidative content from the pulp of *Passiflora liguralis*.JAgri Fd Chem. 55(18):7427-7433. ISSN 0021-8561.

Trillas MI, Casanova E, Cotxarrera L *et al.*, 2006. Composts from agricultural waste and the *Trichoderma asperellum* strain T-34 suppress *Rhizoctonia solani* in cucumber seedlings. BiologicalCont. 39:32-38.

Tsuda T, Horio F, Uchida K, Aoki H and Osawa T. 2003. Dietary cyanidin 3-O-beta-D-glucoside-rich purple corn color prevents obesity and ameliorates hyperglycemia in mice. J. Nutr.133: 2125-2130.

U.S. environmental protection agency. 2012. Local government climate and energy strategy guides. Resource conservation and recovery. A guide to developing and implementing green house gas reduction programs.

UPOV (International Union for the Protection of New Varieties of Plants). 2002.General introduction to the examination of distinctness, uniformity and stability and the development of harmonized descriptions of new varieties of plants. Available at http://www.upov.int/en/publications/tg-rom/tg001/tg_1_3.pdf. Accessed December 12, 2003.

USDA. 1992. Soil Conservation Service. Agricultural Waste Management. Field Handbook.

Vanaja G, Villetchka Gotcheva, Angel Angelov and Renu Agrawal. 2010. Volatile profile of compounds with therapeutic importance produced by probiotic culture strain of *Lb. plantarum* LB cfr.J Fd Sci Technol. 6948 (1):110-113.

Vanitha R and Renu Agrawal. 2012. Purification of protein from probiotic *Leuconostoc mesenteroides* active against *V. cholerae*. Res J Biotechnol. 7 (1): 38-42.

Vastrad BM and Neelagund SE. 2011a. Optimization and production of neomycinfrom diferent agro industrial wastes in solid state fermentation. Intern JPharma Sci Drug Res.3: 104–111.

Vastrad BM and Neelagund SE. 2011b. Optimization of process parameters for rifamycinb production under solid state fermentation from *Amycolatopsis mediterranean* MTCC 14. Intern J Curr Pharm Res. 4:101–108.

Vasudevan N and Rajaram P. 2001. Bioremediation of Oil Sludge-Contaminated Soil. Environ Int 26: 409-411.

Venkateshwaran G, Somashekar D, Prakash MH, Renu Agrawal and Basappa SC.1999. Production and utilization of catalase by using *Saccharomyces cerevisiae* CFR-201.Process Biochem. 34: 187-191.

Venkateswarlu B, Rao AV. 1983. Response of pearl millet to inoculation with different strains of *Azospirillum brasilense*. Plant and Soil. 74 (3): 379-386.

Vicas S I, Bandici L, EusdeaACT and Bandici G E. 2016. Extraction of bioactive compounds from two grape varieties using pulsed electric field.Bulletin of university of agricultural sciences and veterinary medicine cluj-napoca.FdScitechnol.Bulletin UASVM Food Science and Technology 73(1) :85-92.

Vicuña R. 1988. Bacterial degradation of lignin. Review Article. Enz Microb Technol.10 (11): 646-655.

Vidhyalakshmi R, Vallinachiyar C and Radhika R. 2012. Production of xanthan fromagro-industrial waste. J Adv Sci Res 3:56–59.

Vidya C Muralidhar and Renu Agrawal. 2003. Production of verbenol, a high valued food flavourant from improved culture strain of *A.niger* using protoplast fusion.Appl Microbiol Biotechnol. 62 (4): 421-422.

VidyaLaxme B, A Rovetto, R Grau and Renu Agrawal. 2014. Synergistic effects of probiotic *Leuconostoc mesenteroides* and *Bacillus subtilis* in ragi (*Eleucine corocana*) malt for antagonistic activity against *Vibrio cholerae* and other beneficial properties.J Fd Sci Technol. DOI: 10.1007/s13197-012-0834-5.

Vijayaraghavana P, Samuel GP, Vincenta MV and Arasub N AAl-Dhabi. 2016. Bioconversion of agro-industrial wastes for the production of fibrinolytic enzyme from *Bacillus halodurans* IND18: Purification and biochemical characterization. Electronic J Biotechnol. 20: 1-8.

Vijayendra SVN, Renu Agrawal, Taranathan RT and Prasad MS. 2003. Screening and isolation of gel forming exopolysaccharide producing microorganisms from soil samples.Trends in Carbohydrate Chemistry.8 (2): 56-61.

Vijayendra SVN, Vanajakshi V, Varadaraj MC, Venkateswaran Gand Renu Agrawal. 2015. Effect of probiotic cultures on fermentation of Moringa leaves extract. LWT Fd. Sci.Technol. 63(2): 1268-1273.

Vilkhu K, Mawson R, Simons L and Bates D. 2008. Applications and opportunities for ultrasound assisted extraction in the food industry. Innov Fd Sci Emerg. 9: 161-169.

Visioli F and Galli C. 2002. Biological properties of olive oil phytochemicals. Crit Rev Fd Sci Nutr. 42:209–221.

Visioli F, Bogani P and Galli C. 2006. In: Healthful properties of olive oil minor components. Olive Oil- Chemistry and Technology. Boskou D (Ed.) AOCS Press; Champaign, IL: 173–190.

Volf V and Popa I. 2004. The obtaining of active compounds with antioxidant properties from vegetable by-products.Study of the extraction process of polyphenolic compounds from *Vitis* sp. Wood Revista de Chimie.55:707-710.

Wagdy S M and Fakhriya ST. 2012. Primary Assessment of the Biological Activity of *Jojoba* Hull Extracts. Life Sci J. 9(2): 244-253.

Wang Y and Chang Q. 1999. The present conditions of China's iron tailings resource and the art-of-state and effort direction of its utilization.Metal mine. In Chinese.271 (1):1-6.

Warnecke R. 2000. Gasification of biomass: comparison of fixed bed and fluidized bed gasifier. Biomass Bioenergy.18:489–497.Waste Management, McGraw-Hill, Inc. New Delhi.

Widmer W, Zhou W and Grohmann K. 2010. Pretreatment effects on orange processing waste for making ethanol by simultaneous saccharification and fermentation. Biores Tech. 5242-5249.

Wilkins MR, Widmer WW and Grohmann K. 2007. Simultaneous saccharification and fermentation of citrus peel waste by *Saccharomyces cerevisiae* to produce ethanol.Proc Biochem. 42 (12):1614-1619.

Williamson-Hughes P S, Flickinger B D, Messina M J and Mark W. Empie. 2006. Isoflavone supplements containing predominantly genistein reduce hot flash symptoms: a critical review of published studies.Menopause: The Journal of The North American Menopause Society. 13(5): 831-839 DOI: 10.1097/01.gme.0000227330.49081.9e

Williamson G and Manach C. 2005.Bioavailability and bioefficacy of polyphenols in humans. II. Review of 93 intervention studies. The American J Clin Nutri. 81(1): 243S–255S.https://doi.org/10.1093/ajcn/81.1.243S

Wiselogel A, Tyson S and Johnson D. 1996. Biomass feedstock resources and composition. In: Handbook on bioethanol: Production and utilization.Taylor & Francis, Washington, DC. (Ed)Wyman CE.105–118.

WISErg Corporation. Sound Consumer.2015.Converting food waste into organic fertilizer. 15459, E95th St.Redmond, Washington 98052.WISErg's mission is to rethink food waste, converting an urban resource into a nutrient dense fertilizer to grow nutrient rich foods.

Woodley JM and Lilly MD.1994. Biotransformation reactor selection and operation.Applied biocatalysis, Harwood Academic, Chur. (Eds.):J.M.S. Cabral, D. Best, L. Boross, J. Tramper.

Wu D, Zhu S and Wang Z. 2001. Status quo and suggestions on comprehensively utilizing mining tailings at home and abroad. J Qingdao InsArchit Eng. In Chinese. 22(4): 84-87.

Wu GW, Gao JM, Shi XW, Zhang Q, Wei SP, Ding K. 2011. Microbial transformations of diosgenin by the white-rot basidiomycete Coriolus versicolor.J Nat Prod. 28; 74(10):2095-101. doi: 10.1021/np2003484.

Wu T, Schoenau JJ, Li F, Qian P, Malhi SS, *et al.*, 2005. Influence of fertilization and organic amendments on organic carbon fractions in Heilu soil on the loess plateau of China. J Plant Nutr Soil Sci.168:100–107.

Xia M, Ling WH, Ma J, Kitts DD, Zawistowsk J. 2003. Supplementation of diets with black rice pigment fraction attenuates atherosclerotic plaque formation in apolipoprotein E-deficient mice. J Nutr.133:744–51.

Xiao L, Yi J and Cui Z. 2010. Iron tailings comprehensive utilization at home and abroad. J Jilin Ins Architecture Civil Eng. In Chinese.27 (4): 22-26.

Xiao Z, Liu P, Qin JY and Xu P.2007. Statistical optimization of medium components for enhanced acetoin production from molasses and soybean meal hydrolysate. Appl Microbiol Biotechnol.74(1):61–68.

Xing J, Song S and Xu X.2001.Preparation of gold tailings glass-ceramics.The Chinese J Nonferrous Metals.11 (2):319-322.

Xu H, Yiqi Yang.2012. Bioplastics from waste materials and low-value byproduct, degradable polymers and materials: Principles and Practice (2nd Ed.),Chap. 8:113–140. Chapter DOI: 10.1021/bk-2012-1114.ch008,ACSSymposiumSeries,1114.ISBN13:9780841228221eISBN: 9780841228238.

Yan W, Acharjee TC, Coronella CJ and Vasquez VR. 2009. Thermal pretreatment of lignocellulosic biomass. Environ Prog Sustain Energy.28: 435–440.

Yang H, Yan R, Chen H, Lee DH, Liang DT and Zheng C. 2006. Pyrolysis of palm oil wastes for enhanced production of hydrogen rich gases. Fuel Process Technol. 87:935–942.

Ye M, Han J, An D, Tu G and Guo G. 2005. New cytotoxic bufadienolides from the biotransformation of resibufogenin by *Mucor polymorphosporus*.Tetrahedron.61: 8947–8955.

Yeung PY, Johnson RL and Xu JG. 1997. Biodegradation of petroleum hydrocarbons in soil as affected by heating and forced aeration.J Environ Qual. 26 (6):1511-1516.

Zaidon A, Nizam AMN, Nor MYM, Abood F, Paridah MT, Yuziah MYN and Jalaluddin H. 2007. Properties of particle board made from pretreated particles of rubber wood, EFB, and rubberwood–EFB blend. J Appl Sci.7: 1145–1151.

Zargham R and Zargham H. 2008. Tannin extracted from Sumac inhibits vascular smooth muscle cell migration.Mc Gill J. Med.11: 119–123.

Zeisel SH.1999. Regulation of "Nutraceuticals". Sci. 285:185–186.

Zengxiang Lu and Meifeng C. 2012. Study on methods of environmental restoration of a mine in transition from open-pit to underground mining.In: Asia Pacific Conference on Environmental Science and Technology (Kuala Lumpur, Malaysia).USA: Information Engineering Research Institute. Wei Lee (Ed.) 551-555.

Zengxiang Lu. 2012. Disposal Methods on Solid Wastes from Mines in Transition from Open-Pit to Underground Mining. Procedia Environmental Sciences.16:715-721.

Zhan J, Liu W, Guo H, Zhang Y and Guo D. 2003. Selective dehydrogenation of resibufogenin and cinobufagin at 3-OH by *Pseudomonas aeruginosa*. Enzyme Microb Technol. 33:29-32.

Zhang C and Taihua Mu. 2011. Optimisation of pectin extraction from sweet potato (*Ipomoea batatas*, Convolvulaceae) residues with disodium phosphate solution by response surface method. Int J Fd Sci Technol. 1365-2621.

Zhang J, Sun Y, Liu JH, Yu BY and Xu Q. 2007. Microbial transformation of three bufadienolides by *Nocardia* sp. and some insight for the cytotoxic structure–activity relationship (SAR). Bioorg Med Chem Lett.17:6062–6065.

Zhang J, Wen N and Yali W. 2005. Reaserch on producing glass ceramics by iron tailings. Metal Mine. In Chinese, 353(11): 72-74.

Zhang LH, Xu CB and Champagne P. 2010. Overview of recent advances in thermo-chemical conversion of biomass. Energy Convers Manage. 51:969–982.

Zhang S, Xue X, Liu R and Jin Z. 2005. Current situation and prospect of the comprehensive utilization of mining tailings. Mining Metallurgical Eng. 25(3):44-47.

Zheng L, Zheng P, Sun Z, Bai Y, Wang J and Guo X. 2007. Production of vanillin from waste residue of rice bran oil by *Aspergillus niger* and *Pycnoporus cinnabarinus*. Bioresour Technol. 98 (5):1115–1119.

Zheng T, Lidan Y, Yuling Z and Zhao B. 2014. Evaluation of different pretreatments on microbial transformation of saponins in *Dioscorea zingiberensis* for diosgenin production. Biotechnol Equip. 28(4):740–746.

Zhou L. 2007. Investigation and practice on mining land rehabilitation and ecological reconstruction in China. Nonferrous Metals. 59 (2):90-94.

Zhu C, Yang G, Li H, Du D and Lin Y.2015. Electrochemical sensors and biosensors based on nanomaterials and nanostructures. Anal. Chem. 87(1): 230–249.

Zimnoch G E, Marczewski W, Lebecka R, Flis B, Schafer P R, Salamini F and Gebhardt C. 2000. QTL analysis of new sources of resistance to *Erwinia carotovora* ssp. atroseptica in potato done by AFLP, RFLP and resistance-gene-like markers. Crop Sci. 40:1156–1167.

FSC
www.fsc.org
MIX
Papier | Fördert gute Waldnutzung
FSC® C083411

Kolibri 360 GmbH
Ferdinand-Jühlke-Straße 7
99095 Erfurt, Deutschland
produktsicherheit@kolibri360.de